Promising Research and Strategies in Wastewater Treatment, Sludge Management and Valorization

Promising Research and Strategies in Wastewater Treatment, Sludge Management and Valorization

Editors

Amanda Laca Pérez
Yolanda Patiño

Basel • Beijing • Wuhan • Barcelona • Belgrade • Novi Sad • Cluj • Manchester

Editors
Amanda Laca Pérez
Chemical and
Environmental Engineering
University of Oviedo
Oviedo, Spain

Yolanda Patiño
Chemical and
Environmental Engineering
University of Oviedo
Oviedo, Spain

Editorial Office
MDPI
St. Alban-Anlage 66
4052 Basel, Switzerland

This is a reprint of articles from the Special Issue published online in the open access journal *Applied Sciences* (ISSN 2076-3417) (available at: https://www.mdpi.com/journal/applsci/special_issues/Wastewater_Sludge_Management).

For citation purposes, cite each article independently as indicated on the article page online and as indicated below:

Lastname, A.A.; Lastname, B.B. Article Title. *Journal Name* **Year**, *Volume Number*, Page Range.

ISBN 978-3-0365-9114-8 (Hbk)
ISBN 978-3-0365-9115-5 (PDF)
doi.org/10.3390/books978-3-0365-9115-5

Cover image courtesy of Ana Isabel Díaz González

© 2023 by the authors. Articles in this book are Open Access and distributed under the Creative Commons Attribution (CC BY) license. The book as a whole is distributed by MDPI under the terms and conditions of the Creative Commons Attribution-NonCommercial-NoDerivs (CC BY-NC-ND) license.

Contents

About the Editors vii

Amanda Laca and Yolanda Patiño
Special Issue on Promising Research and Strategies in Wastewater Treatment, Sludge Management, and Valorisation: Volume I
Reprinted from: *Appl. Sci.* **2023**, *13*, 10121, doi:10.3390/app131810121 1

Makoto Shigei, Almoayied Assayed, Ayat Hazaymeh and Sahar S. Dalahmeh
Pharmaceutical and Antibiotic Pollutant Levels in Wastewater and the Waters of the Zarqa River, Jordan
Reprinted from: *Appl. Sci.* **2021**, *11*, 8638, doi:10.3390/app11188638 5

Andrea Menéndez-Manjón, Reyes Martínez-Díez, Daniel Sol, Amanda Laca, Adriana Laca, Amador Rancaño and Mario Díaz
Long-Term Occurrence and Fate of Microplastics in WWTPs: A Case Study in Southwest Europe
Reprinted from: *Appl. Sci.* **2022**, *12*, 2133, doi:10.3390/app12042133 17

Matthew Mamera, Johan J. van Tol, Makhosazana P. Aghoghovwia, Alfredo B. J. C. Nhantumbo, Lydia M. Chabala, Armindo Cambule, et al.
Potential Use of Biochar in Pit Latrines as a Faecal Sludge Management Strategy to Reduce Water Resource Contaminations: A Review
Reprinted from: *Appl. Sci.* **2021**, *11*, 11772, doi:10.3390/app112411772 35

Badr Bouhcain, Daniela Carrillo-Peña, Fouad El Mansouri, Yassine Ez Zoubi, Raúl Mateos, Antonio Morán, et al.
Removal of Emerging Contaminants as Diclofenac and Caffeine Using Activated Carbon Obtained from Argan Fruit Shells
Reprinted from: *Appl. Sci.* **2022**, *12*, 2922, doi:10.3390/app12062922 53

Brahim Samir, Nabil Bouazizi, Patrick Nkuigue Fotsing, Julie Cosme, Veronique Marquis, Guilherme Luiz Dotto, et al.
Preparation and Modification of Activated Carbon for the Removal of Pharmaceutical Compounds via Adsorption and Photodegradation Processes: A Comparative Study
Reprinted from: *Appl. Sci.* **2023**, *13*, 8074, doi:10.3390/app13148074 73

Justyna Staninska-Pieta, Jakub Czarny, Wojciech Juzwa, Łukasz Wolko, Paweł Cyplik and Agnieszka Piotrowska-Cyplik
Dose–Response Effect of Nitrogen on Microbial Community during Hydrocarbon Biodegradation in Simplified Model System
Reprinted from: *Appl. Sci.* **2022**, *12*, 6012, doi:10.3390/app12126012 87

Ana Isabel Díaz, Marta Ibañez, Adriana Laca and Mario Díaz
Biodegradation of Olive Mill Effluent by White-Rot Fungi
Reprinted from: *Appl. Sci.* **2021**, *11*, 9930, doi:10.3390/app11219930 101

Manasik M. Nour, Maha A. Tony and Hossam A. Nabwey
Heterogeneous Fenton Oxidation with Natural Clay for Textile Levafix Dark Blue Dye Removal from Aqueous Effluent
Reprinted from: *Appl. Sci.* **2023**, *13*, 8948, doi:10.3390/app13158948 113

About the Editors

Amanda Laca Pérez

Amanda Laca Pérez obtained a bachelor´s degree in biology from the University of Oviedo in 2002. She completed her training with a master's degree in food biotechnology from the University of Oviedo (2004–2006) and with different specialization courses. In addition, she finished her PhD degree in the Department of Chemical and Environmental Engineering of the same University in 2010.

She has published more than sixty book chapters and scientific papers in specialised journals, fifty of which were indexed in the JCR (Journal Citation Reports), and is a coauthor of multiple papers presented in national and international scientific congresses and conferences. She has been a member of the research team of different projects and a main researcher under several contracts with different companies.

Regarding her professional activity, she carried out internships in different companies and worked as Quality Manager. She was also a laboratory technician for three years at the Technical Scientific Services of the University of Oviedo. Since 2019, she has been Assistant Professor in the field of environmental technologies at this university.

Yolanda Patiño

Yolanda Patiño received a BSc degree in chemical engineering and her PhD degree from the University of Oviedo, Spain, in 2010 and 2016, respectively. She is currently Assistant Professor in the Department of Chemical and Environmental Engineering at the University of Oviedo, where she teaches a variety of courses in environmental engineering. Her research in the University of Oviedo was complemented with one international stay at the University of Antwerpen in 2014. Her research interest include some of the following: adsorption processes, the electrochemical removal of emerging pollutants and waste biomass valorization for the obtention of fuels.

Editorial

Special Issue on Promising Research and Strategies in Wastewater Treatment, Sludge Management, and Valorisation: Volume I

Amanda Laca * and Yolanda Patiño *

Department of Chemical and Environmental Engineering, University of Oviedo, C/Julián Clavería s/n, 33006 Oviedo, Spain
* Correspondence: lacaamanda@uniovi.es (A.L.); patinoyolanda@uniovi.es (Y.P.)

Citation: Laca, A.; Patiño, Y. Special Issue on Promising Research and Strategies in Wastewater Treatment, Sludge Management, and Valorisation: Volume I. *Appl. Sci.* **2023**, *13*, 10121. https://doi.org/10.3390/app131810121

Received: 4 September 2023
Accepted: 7 September 2023
Published: 8 September 2023

Copyright: © 2023 by the authors. Licensee MDPI, Basel, Switzerland. This article is an open access article distributed under the terms and conditions of the Creative Commons Attribution (CC BY) license (https://creativecommons.org/licenses/by/4.0/).

1. Introduction

Rapid urbanization and industrialization, together with new contaminants arising from many different sources, make it necessary to move forwards with research to face future challenges regarding water pollution.

Conventional wastewater treatment plants (WWTPs) are designed primarily to remove organic matter and nutrients. For this reason, in many cases, these treatments are inefficient for the removal of specific pollutants, with the consequent risk that this entails.

Emerging contaminants (ECs) represent a wide group of potentially hazardous compounds that have been classified into various categories, including pharmaceutical and personal care products (PPCPs), per- and poly-fluoroalkyl substances (PFAS), flame retardants, surfactants, endocrine-disrupting chemicals (EDCs), and microplastics (MPs) [1]. Recently, it has been highlighted that chemical and plastic pollution has surpassed global boundaries, resulting in harmful effects, not only for the environment, but also for the humanity [2].

Furthermore, large amounts of sludge—the main residue originating from WWTPs—are produced every year. The management of this waste leads to high costs, both in economic and environmental terms, so exploring alternatives for its valorisation is essential. Residue reduction and reuse of wastewater seem to be an excellent option to achieve a circular economy, lessening the environmental impacts associated with wastewater treatment processes. In addition, different efforts to valorise agricultural by-products to treat wastewater have been conducted, which also meets sustainability goals.

In view of the current situation, this Special Issue aimed to compile the latest research on relevant wastewater treatment concerns, such as EC occurrence and removal in WWTPs, novel technologies for wastewater and sludge treatment, sludge valorisation, transformation of agricultural residues into materials for wastewater treatment, etc.

Of the papers submitted to this Special Issue, eight were finally accepted and published. These articles can be organized into three different topics, as synthetized below.

2. Occurrence and Fate of Emerging Contaminants in WWTPs

WWTPs have been reported as one of the main sources for the release of ECs into the environment. Accordingly, gaining an in depth knowledge of the occurrence and fate of these potentially harmful contaminants is a topic of great interest. In this context, Kumar et al. [3] studied the incidence of 15 pharmaceutically active compounds (atenolol, amlodipine, bisoprolol, carbamazepine, citalopram, diazinon, fluoxetine, ketoconazole, metformin, metoprolol, oxazepam, paracetamol, propranolol, risperidone, and sertraline), 18 antibiotics (ciprofloxacin, clarithromycin, clindamycin, doxycycline, erythromycin, ofloxacin, linezolid, metronidazole, moxifloxacin, norfloxacin, tetracycline, trimethoprim, amoxicillin, ampicillin, benzylpenicillin, fusidic acid, rifampicin, and sulfamethoxazole), and one stimulant (caffeine) in the largest treatment facility in Jordan (Assamra WWTP). These authors also analysed the presence of these compounds in the Zarqa River, where the treated water is discharged. This study concluded that Assamra WWTP is efficient in removing caffeine and pharmaceutically active compounds (except for bisoprolol and carbamazepine) with overall efficiencies higher than 80%, whereas the wastewater treatment process was not able to eliminate antibiotics. Additionally, Zarqa river water was shown to be contaminated with pharmaceutically active compounds and antibiotics, and the origins of this pollution were the effluent discharges from Assamra WWTP and side-inputs from the areas surrounding the river.

Menéndez-Manjón et al. [4] remarked that WWTPs also represent a major indirect source of microplastics released into the environment and analysed, as a case study, the performance of a WWTP sited in Southwest Europe for one year. They observed that the majority of MPs detected in wastewater and sludge samples were fragments and fibres. Regarding the chemical composition of these micropollutants in the water samples, polyethylene (PE), polyethylene terephthalate (PET), and polypropylene (PP) were the most common microplastics, whereas, in the sludge samples, the main polymers were PET, polyamide (PA), and polystyrene (PS). No significant variations were found between months and the results showed that removal efficiencies were between 89% and 95% during study the period. Moreover, most MPs (88%) were eliminated in the secondary treatment stage, being entrapped in the sludge.

3. Agricultural Residue Valorisation for Wastewater Treatment

It has been estimated that approximately 998 million tons of agricultural wastes are generated each year. In the search for economic and eco-friendly alternatives to management, agricultural residues have been proposed as biosorption materials for the removal of water pollutants [5].

Mamera et al. [6] reviewed the literature published on the potential use of biochar, a carbon-rich adsorbent produced from different organic biomass, in faecal sludge management in developing countries. This work determined that biochar is a viable option for faecal sludge management due to its capacity to bind different inorganic and organic pollutants. Incorporating biochar as a low-cost adsorbent in pit latrine sludge management could lead an improvement in the quality of water resources and, in addition, biochar-amended sludge could be repurposed as a useful economical by-product.

Bouhcain et al. [7] obtained activated carbon from argan nutshells by means of chemical activation. Once this material was characterized, it was assayed as an adsorbent for the removal of two emerging contaminants employed as model pollutants, a stimulant (caffeine) and an anti-inflammatory drug (diclofenac). The highest adsorption capacity was about 126 mg and 210 mg per gram of activated carbon, for diclofenac and caffeine, respectively. The adsorption process was described by a pseudo-second-order kinetic model and the thermodynamic parameters indicated that this process was spontaneous and exothermic for diclofenac and endothermic in the case of caffeine.

Samir et al. [8] employed date stems as a precursor to prepare activated carbon (AC) by calcination. The AC was modified by a hydroxylation strategy to increase the hydroxyl groups over the surface, resulting in AC-OH. In addition, to ensure that photodegradation took place, AC was impregnated into TiO_2 solution to produce AC-TiO_2. The obtained materials were evaluated to remove pharmaceutical contaminants based on atenolol (AT) and propranolol (PR). Results showed that the removal of AT and PR reached 92% by adsorption, while 94% was obtained by photodegradation. Comparing both processes, adsorption proved to be more suitable for removing pollutants from water, since it presented low energy consumption, which revealed AC-OH as a low-cost and environmentally friendly material suitable for wastewater treatment on an industrial scale.

4. Removal of Recalcitrant Compounds

Biodegradation is a cost-effective and practical solution for removing contaminants from different environments, including wastewater, where microorganisms play a key role. However, biodegradation of recalcitrant pollutants is particularly problematic due to the lack of efficient microbial metabolic traits [9]. A fundamental aspect of biodegradation is the C:N ratio. Stanińska-Pięta et al. [10] assessed the impact of nitrogen compounds during the process of biological decomposition of hydrocarbons, confirming the positive effect of properly optimised biostimulation. Nevertheless, when excessive biostimulation was employed, negative effects on the biodegradation efficiency were observed. Certain effluents from the food industry, such as the olive sector, contain lignocellulosic organic matter and phenolic compounds, which are difficult to eliminate using conventional biological methods. Díaz et al. [11] evaluated *Phanerochaete chrysosporium* to treat "alperujo" (olive pomace), a by-product composed of pulp waste, ground stone, and skin, together with vegetation waters. COD (Chemical Oxygen Demand) and colour removals of around 60%, and 32% of total phenolic compounds degradation, were achieved in this work, showing the interest of this fungi in recalcitrant compound treatment.

Furthermore, Fuller's earth was proposed as an environmentally green material to be employed as a catalyst in heterogeneous Fenton oxidation technology [12]. Fuller's earth was chemically and thermally activated, and the obtained catalysts were employed to treat synthetic wastewater polluted with Levafix Dark Blue dye. Optimal results were observed when 818 and 1.02 mg/L of Fuller's earth and hydrogen peroxide were used, respectively. Specifically, it was possible to achieve a removal efficiency of 99% and, in addition, after a six-cycle test, a reasonable percentage of dye (73%) was still removed. This underlines the potential of this material to be applied in textile wastewater effluent treatment.

5. Future Prospects

Approximately, 80% of the world's wastewater is still discharged into the environment untreated. With an increasing scarcity of freshwater available, mainly due to a growing population and the unsustainable use of natural resources, the treatment and recycling of wastewater is a topic of great interest. Moreover, the occurrence of emerging pollutants in wastewaters, such as pathogens, pharmaceuticals, and microplastics, has become a progressively serious issue. Although conventional contaminants may be feasibly removed using established methods, many hazardous compounds, for example, polyfluoroalkyl substances, antibiotics, endocrine-disrupting chemicals, etc., are resistant to typical biological treatments, which leads to a severe threat, not only for aquatic environments, but also for human health.

According to the United Nations, it is very likely that water will be the most critical natural resource in the decades to come. Certainly, universal access to clean water and sanitation is one of the 17 Sustainable Development Goals (SDG 6) that should be achieved by 2030 [13]. This makes it essential to develop novel green alternatives to address wastewater purification in the present context of sustainability.

Moreover, following circular economy principles, valorisation of wastes is mandatory today; this requires, for instance, the use of agricultural residues in wastewater treatment processes, the use of sewage sludge as soil amendment, or the recovery of compounds of interest from different industrial wastes.

Thus, the current defiance of the wastewater sector could be summarised as the need for sustainable and cost-effective technology development. More efforts are required from the scientific community to tackle scientific, political, and societal aspects regarding wastewater concerns in the context of demanding environmental conditions and challenges of the future.

Author Contributions: Conceptualization, formal analysis, writing—original draft preparation, writing—review and editing, A.L. and Y.P. All authors have read and agreed to the published version of the manuscript.

Acknowledgments: All the authors and peer reviewers are gratefully thanked for their valuable contributions that made this Special Issue possible. The MDPI is congratulated for their dedicated editorial support. Finally, we want to place on record our sincere gratefulness to the Assistant Editor, who kindly supported us during the development of this project.

Conflicts of Interest: The authors declare no conflict of interest.

References

1. Kumar, R.; Vuppaladadiyam, A.K.; Antunes, E.; Whelan, A.; Fearon, R.; Sheehan, M.; Reeves, L. Emerging Contaminants in Biosolids: Presence, Fate and Analytical Techniques. *Emerg. Contam.* **2022**, *8*, 162–194. [CrossRef]
2. Yang, L.; Weber, R.; Liu, G. Science and policy of legacy and emerging POPs towards Implementing International Treaties. *Emerg. Contam.* **2022**, *8*, 299–303. [CrossRef]
3. Shigei, M.; Assayed, A.; Hazaymeh, A.; Dalahmeh, S.S. Pharmaceutical and Antibiotic Pollutant Levels in Wastewater and the Waters of the Zarqa River, Jordan. *Appl. Sci.* **2021**, *11*, 8638. [CrossRef]
4. Menéndez-Manjón, A.; Martínez-Díez, R.; Sol, D.; Laca, A.; Laca, A.; Rancaño, A.; Díaz, M. Long-Term Occurrence and Fate of Microplastics in WWTPs: A Case Study in Southwest Europe. *Appl. Sci.* **2022**, *12*, 2133. [CrossRef]
5. Karić, N.; Maia, A.S.; Teodorović, A.; Atanasova, N.; Langergraber, G.; Crini, G.; Ribeiro, A.R.L.; Ðolić, M. Bio-waste Valorisation: Agricultural Wastes as Biosorbents for Removal of (In)organic pollutants in Wastewater Treatment. *CEJ Adv.* **2022**, *9*, 100239. [CrossRef]
6. Mamera, M.; van Tol, J.J.; Aghoghovwia, M.P.; Nhantumbo, A.B.J.C.; Chabala, L.M.; Cambule, A.; Chalwe, H.; Mufume, J.C.; Rafael, R.B.A. Potential Use of Biochar in Pit Latrines as a Faecal Sludge Management Strategy to Reduce Water Resource Contaminations: A Review. *Appl. Sci.* **2021**, *11*, 11772. [CrossRef]
7. Bouhcain, B.; Carrillo-Peña, D.; El Mansouri, F.; Ez Zoubi, Y.; Mateos, R.; Morán, A.; Quiroga, J.M.; Zerrouk, M.H. Removal of Emerging Contaminants as Diclofenac and Caffeine Using Activated Carbon Obtained from Argan Fruit Shells. *Appl. Sci.* **2022**, *12*, 2922. [CrossRef]
8. Samir, B.; Bouazizi, N.; Nkuigue Fotsing, P.; Cosme, J.; Marquis, V.; Dotto, G.L.; Le Derf, F.; Vieillard, J. Preparation and Modification of Activated Carbon for the Removal of Pharmaceutical Compounds via Adsorption and Photodegradation Processes: A Comparative Study. *Appl. Sci.* **2023**, *13*, 8074. [CrossRef]
9. Bala, S.; Garg, D.; Thirumalesh, B.V.; Sharma, M.; Sridhar, K.; Inbaraj, B.S.; Tripathi, M. Recent Strategies for Bioremediation of Emerging Pollutants: A Review for a Green and Sustainable Environment. *Toxics* **2022**, *10*, 484. [CrossRef] [PubMed]
10. Staninska-Pięta, J.; Czarny, J.; Juzwa, W.; Wolko, Ł.; Cyplik, P.; Piotrowska-Cyplik, A. Dose–Response Effect of Nitrogen on Microbial Community during Hydrocarbon Biodegradation in Simplified Model System. *Appl. Sci.* **2022**, *12*, 6012. [CrossRef]
11. Díaz, A.I.; Ibañez, M.; Laca, A.; Díaz, M. Biodegradation of Olive Mill Effluent by White-Rot Fungi. *Appl. Sci.* **2021**, *11*, 9930. [CrossRef]
12. Nour, M.M.; Tony, M.A.; Nabwey, H.A. Heterogeneous Fenton Oxidation with Natural Clay for Textile Levafix Dark Blue Dye Removal from Aqueous Effluent. *Appl. Sci.* **2023**, *13*, 8548. [CrossRef]
13. United Nations. Available online: https://www.un.org/en/ (accessed on 26 July 2023).

Disclaimer/Publisher's Note: The statements, opinions and data contained in all publications are solely those of the individual author(s) and contributor(s) and not of MDPI and/or the editor(s). MDPI and/or the editor(s) disclaim responsibility for any injury to people or property resulting from any ideas, methods, instructions or products referred to in the content.

Article

Pharmaceutical and Antibiotic Pollutant Levels in Wastewater and the Waters of the Zarqa River, Jordan

Makoto Shigei [1], Almoayied Assayed [2], Ayat Hazaymeh [2] and Sahar S. Dalahmeh [1,*]

[1] Department of Earth Sciences, Uppsala University, P.O. Box Villavägen 16, SE 753 38 Uppsala, Sweden; makoto.shigei@geo.uu.se
[2] Royal Scientific Society, P.O. Box 1438, Amman 11941, Jordan; Almoayied.Assayed@rss.jo (A.A.); Ayat.Hazaymeh@rss.jo (A.H.)
* Correspondence: Sahar.Dalahmeh@geo.uu.se

Abstract: Assamra wastewater treatment plant (WWTP) is the largest treatment facility in Jordan. Treated wastewater is discharged into the Zarqa River (ZR) and used to irrigate fodder and vegetables. ZR also includes surface runoff, stormwater, and raw wastewater illegally discharged into the river. This study examined pharmaceutically active compounds (PhAC) in water resources in the ZR basin. Samples of WWTP influent and effluent and river water from four sites along ZR were collected. Concentrations of 18 target antibiotics, one stimulant, and 15 other PhACs were determined in the samples. Five antibiotics were detected in WWTP influent (510–860 ng L^{-1} for $\sum$Antibiotics) and six in the effluent (2300–2600 ng L^{-1} for $\sum$Antibiotics). Concentrations in the effluent of all antibiotics except clarithromycin increased by 2- to 5-fold compared with those in influent, while clarithromycin concentration decreased by around 4- fold (from 308 to 82 ng L^{-1}). WWTP influent and effluent samples contained 14 non-antibiotic PhACs, one simulant, and six antibiotics at detectable concentrations. The dominant PhACs were paracetamol (74% of $\sum$PhACs) in the influent and carbamazepine (78% of $\sum$PhACs) in the effluent. At ZR sampling sites, carbamazepine was the dominant PhAC in all cases (800–2700 ng L^{-1}). The antibiotics detected in WWTP effluent were also detected at the ZR sites. In summary, water in ZR is contaminated with PhACs, including antibiotics, and wastewater discharge seems to be the main pathway for this contamination. The occurrence of antibiotics and other PhACs in the irrigated soil requires investigation to assess their fate.

Keywords: Assamra WWTP; caffeine; carbamazepine; irrigation; ofloxacin; paracetamol; pharmaceuticals; Zarqa River

Citation: Shigei, M.; Assayed, A.; Hazaymeh, A.; Dalahmeh, S.S. Pharmaceutical and Antibiotic Pollutant Levels in Wastewater and the Waters of the Zarqa River, Jordan. *Appl. Sci.* **2021**, *11*, 8638. https://doi.org/10.3390/app11188638

Academic Editors: Amanda Laca Pérez and Yolanda Patiño

Received: 6 August 2021
Accepted: 4 September 2021
Published: 17 September 2021

Publisher's Note: MDPI stays neutral with regard to jurisdictional claims in published maps and institutional affiliations.

Copyright: © 2021 by the authors. Licensee MDPI, Basel, Switzerland. This article is an open access article distributed under the terms and conditions of the Creative Commons Attribution (CC BY) license (https://creativecommons.org/licenses/by/4.0/).

1. Introduction

Worldwide, there is large-scale production and use of a vast range of pharmaceutically active compounds (PhACs), including antibiotics. Different regions of the world have different levels of restriction on prescription and sales of drugs. In Jordan, self-medication is common practice, and drugs can easily be purchased from drugstores without prior prescription, despite laws prohibiting the sale and dispensing of non-prescribed antibiotics [1,2]. In the study by Almaaytah et al. [2], more than 70% of drugstores investigated dispensed antibiotics, without prescription, for medical issues that included sore throat, urinary tract infection, diarrhea, and otitis media. Antibiotic resistance genes in different types of bacteria have been reported in isolates from the human population in Jordan [3,4].

After ingestion, PhACs (including antibiotics, stimulants, and illicit drugs) and their metabolites end in human excreta (urine and faeces) and reach the environment via direct discharge or discharge of treated effluents from municipal wastewater systems [5,6]. Many low and middle-income countries host pharmaceutical industries that produce wastewater, which often receives poor treatment, e.g., ending up in the environment or discharged into municipal sewage systems [7–9]. For effective removal of PhACs from wastewater, tertiary treatment steps involving nanotechnologies, adsorption, membrane technologies, or advanced oxidation processes (UV, H_2O_2, photooxidation) are needed [10–13]. These technologies are generally expensive and demand significant resources for maintenance and operation. In many parts of the world, wastewater treatment facilities are overloaded, compromising the treatment efficiency, or not applying a tertiary treatment [14–17]. Consequently, effluents from WWTP constitute a significant source of PhACs, which lead to pollution of water resources, e.g., surface, ground, and lake water upon discharge.

Pollution of water resources with PhACs has been reported around the world. Still, most of the research focusing on these pollutants had been conducted in industrial and high-income countries, e.g., Japan, Europe, and the USA [9,18–24].

In Jordan, a middle-income country in the Mediterranean, wastewater treatment often does not include a tertiary step, and drug prescriptions and sales regulations are not regulated strictly. According to the Ministry of Water and Irrigation, 29 wastewater treatment plants (WWTPs) operate in Jordanian cities, with an estimated annual treated discharge of around 120 Mm^3 of wastewater [25]. The Assamra plant is the largest of Jordan's 29 WWTPs, treating wastewater from more than two million people, mainly in the Amman and Zarqa Governorates [26]. Government and non-government agencies in Jordan are currently promoting the reuse of treated wastewater to mitigate the chronic water shortage in the country and for nutrient recovery, i.e., recycling of phosphorus from sewage back to arable land. As a result, more than 92% of the treated wastewater produced in the main cities in Jordan, i.e., Amman and Zarqa, is used for irrigation [25], mainly in the Zarqa River basin.

The Zarqa River (ZR) is an ecosystem component of great importance for food supply and socioeconomic development in Jordan, as the river water is used to irrigate a wide range of vegetables, fodder crops, and industrial/cash crops in surrounding fields and gardens. The remaining ZR water flows down into King Talal Reservoir, a major water reservoir feeding King Abdullah Canal, from which water is taken for irrigation in the lower Jordan valley [26]. The annual average flow rate in ZR is around 360,000 m^3 day^{-1}. Concerning water sources, ZR receives more than 325,000 m^3 day^{-1} of treated effluent from Assamra WWTP and surface water from Amman, Zarqa, Jerash, and parts of Mafraq [26]. Due to the large amounts of wastewater effluent discharged into ZR, the river can be assumed to be a significant pathway for spreading PhACs into the environment through its use as a source of irrigation water. Pollution with several types of micropollutants, including pesticides and pharmaceuticals, in different water sources in the lower Jordan River has been reported by Tiehm et al. [27], Tiehm et al. [28], and Zemann et al. [29]. A recent study detected 14 PhACs in influent and effluent of Assamra WWTP [30]. However, the pollution loads and fate of PhACs in river water along ZR have not been sufficiently explored, and more research is needed in this region.

This study aimed to determine the PhACs pollution of water resources in Jordan's ZR basin, an example of a low-middle income country. Specific objectives were (i) to determine the occurrence and concentrations of 33 multiclass PhACs (e.g., anti-inflammatory, beta-blockers, antibiotics, anti-diabetics, heart and vascular disease drugs, anti-epileptics, stimulants, and anti-fungal) in wastewater and water resources feeding ZR; and (ii) to investigate the contribution of Assamra WWTP to PhAC levels in ZR water and assess the performance of the WWTP in removing selected PhACs, including antibiotics.

2. Materials and Methods

2.1. Description of the Study Area

Wastewater and water resources contributing to the flow in ZR were studied. A catchment area of 4120 km^2 located in the north of Jordan contributes to the natural streamflow in ZR [31]. The ZR Basin has an arid climate in the east and the southeast. In contrast, the western parts have typical Mediterranean climates that are semiarid in Amman (Capital of Jordan) and dry sub-humid in Ajloun, where rainfall exceeds 560 mm. The western parts are mountainous and characterized by cool temperatures in winter and mild temperatures in summer. The annual rainfall ranges from more than 500 mm in the northwest to less than 100 mm in the east, with an average annual precipitation of 250 mm [32]. The basin hosts 60% of Jordan's population and 85% of all industries in Jordan, and its flood plain is used for agriculture [32]. The water downstream in ZR includes treated effluent from Assamra WWTP and surface runoff and stormwater generated during the rainy season (December–April) [33]. Assamra WWTP treats an average of 365,000 m^3 day^{-1} of municipal wastewater and industrial wastewater for the population of 2,270,000–3,300,000 PE [33,34]. The primary use of ZR water is irrigation fodder and vegetable crops in fields within the ZR flood plain [32]. This study area was also investigated in our previous study, and more details of the area can be found in Shigei et al. [35].

2.2. Sample Collection

In a single sampling event, wastewater and river water samples were collected in the ZR catchment area (Figure 1). The samples of influent and effluent were collected at Assamra WWTP (n = 4). River water samples were collected manually from the top 30 cm water layer of the river by filling high-density polyethylene (HDPE) plastic bottles from the center of the river at four locations: (i) Sukhna station (5.45 km from the main ZR), in a tributary unaffected by Assamra WWTP (n = 2); (ii) Twahin Eledwan station (28.74 km from Assamra WWTP) (n = 2) and (iii) Military station (47.73 km from Assamra WWTP) (n = 2), both downstream of Assamra WWTP; and (iv) Jerash stream, a groundwater stream feeding into ZR (n = 2) (Figure 1, Table S1 in Supporting Information (SI)). Two samples were collected from each site, with an interval of 1 h between the samples. All samples were kept frozen at −20 °C and transported to Sweden for analysis of PhACs.

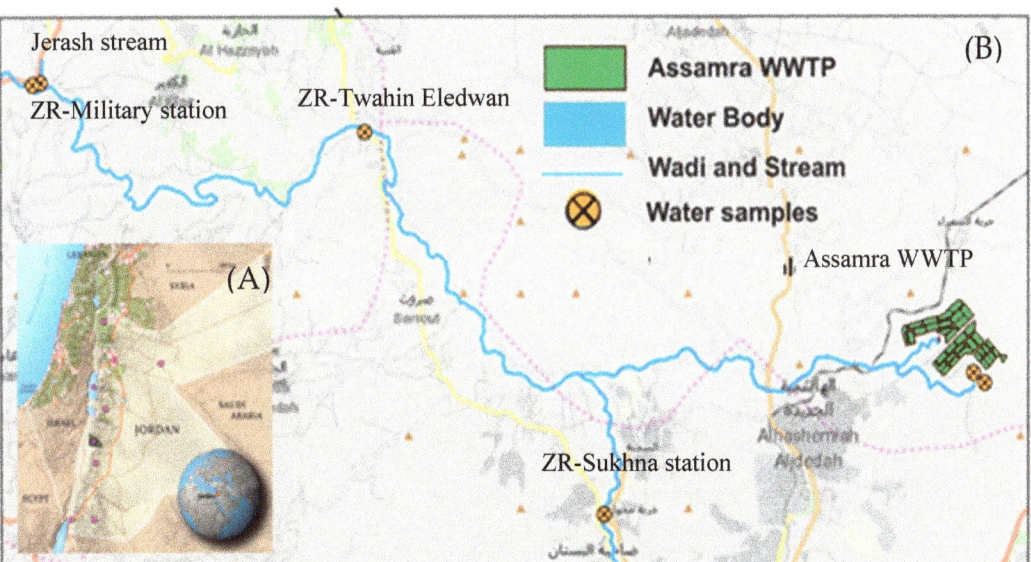

Figure 1. (**A**) Location of the Zarqa River (ZR) basin in Jordan. (**B**) Sampling sites for Assamra wastewater treatment plant (WWTP) influent and effluent (n = 4) and for river water at Sukhna station (n = 2), Twahin Eledwan station (n = 2), Military station (n = 2), and Jerash stream (n = 2). Part A of the diagram is modified from Shigei et al. (2020).

2.3. PhACs Target Analyses

The term PhACs is used hereafter to refer to all substances which have therapeutic effects other than antibiotics. Antibiotics were considered separately because of potential health and environmental impacts. A total of 15 PhACs were included in the analysis (atenolol, amlodipine, bisoprolol, carbamazepine, citalopram, diazinon, fluoxetine, ketoconazole, metformin, metoprolol, oxazepam, paracetamol, propranolol, risperidone, and sertraline). In addition, the concentrations of 18 antibiotics (ciprofloxacin, clarithromycin, clindamycin, doxycycline, erythromycin, ofloxacin, linezolid, metronidazole, moxifloxacin, norfloxacin, tetracycline, trimethoprim, amoxicillin, ampicillin, benzylpenicillin, fusidic acid, rifampicin, and sulfamethoxazole) and one stimulant (caffeine) were analysed. The chemical properties of each compound are shown in Table S1. Isotopically labelled internal standards (IS) used in the analyses were diclofenac $^{13}C_6$, hydrochlorothiazide $^{13}C_6$, carbamazepine $^{13}C^{15}N$, and ibuprofen-d3.

2.4. Extraction and Analyses of PhACs and Other Parameters

All influent, effluent, and river water samples were extracted using solid-phase extraction with Oasis HLB cartridges (200 mg, 6 cc; Waters Corporation, Manchester, UK) according to the extractions method previously described in Dalahmeh et al. [36], Dalahmeh et al. [37], nd Gros et al. [38]. Before extraction, the samples were spiked with 100 µL of IS mixture containing 1 ng µL^{-1} diclofenac $^{13}C_6$, hydrochlorothiazide $^{13}C_6$, carbamazepine $^{13}C^{15}N$, and ibuprofen-d$_3$.

The mass of the target PhACs, antibiotics, and caffeine compounds was determined using high-performance liquid chromatography coupled with mass spectrophotometry (HPLC-MS/MS). All analyses were carried out at the Swedish Environmental Institute (IVL) laboratories using a binary Shimadzu AD20 UFLC HPLC system with automatic sample changer and column furnace coupled to an ABSciex API-4000 mass spectrometer. Samples were analysed under positive and negative electrospray ionization (ESI) mode using a Waters XBridge BEH C18 column (100 mm × 2.1 mm with 3 µm opening size). The eluents used in the mobile phase were A: 10 mM acetic acid in deionized water and B: methanol. The gradient used was a linear gradient from 0–90% methanol for 17 min, with a final plateau at 90% methanol for 4 min before a rapid return to 100% A and a final recovery and equalization of 2 min. The concentration of each analyte was quantified using an eight-point calibration curve (0, 5, 10, 20, 50, 100, 200, and 500 ng).

Besides the target PhACs, antibiotics, and caffeine, the river water, influent and effluent samples were analysed for pH, chemical oxygen demand (COD), electrical conductivity (EC), total suspended solids (TSS), and total solids (TS). All analyses of liquid samples were performed according to Standard Methods for Examination of Water and Wastewater APHA [39], using the following protocols: pH (4500-H and B), biochemical oxygen demand (BOD5; 5210-B), TSS, and TS (2540-B-D). The pH was measured by an electrode that measures the concentration of H ion by millivolts. The Chemical Oxygen Demand was measured by oxidizing the water sample by oxidizing agent (potassium dichromate) followed by open reflux digestion at 1500 °C for 2 h, then back titration for the remaining dichromate using sodium thiosulfate. The electrical conductivity of water was measured using a conductivity cell immersed in a 50 mL sample. Total solid was measured by gravimetric method, through drying the sample in an oven at 105 °C overnight until the crucible has a constant weight. Then, the difference in weights showed the total solids that exist in a sample (summation of total dissolved solids and total suspended solids). Total suspended solid was also measured by gravimetric method, by weighing the washed dried filter paper, then filtering the sample and drying it in an oven at 105 °C, the difference showing the concentration of suspended solids.

2.5. Quality Control

Method blank was prepared using 500 mL of pure MilliQ®® water spiked with 100 µL IS. The blank sample was extracted following the same procedure as used for the other liquid samples. The blank sample did not show detectable levels of any of the measured substances. All detectable concentrations lower than 1 ng L^{-1} were recorded as <1 ng L^{-1}.

2.6. Calculations and Statistical Analysis

Analysis of variance (ANOVA) at 95% confidence level was used to assess the significance of the difference in PhAC concentrations between the different locations. All statistical analyses were performed using the ANOVA adds-in package coupled to Excel 2016 (Microsoft Office, Microsoft, USA).

3. Results and Discussion

3.1. General Quality of Wastewater and Water Resources within Zarqa River Basin

The influent to Assamra WWTP contained 1550 mg L^{-1} of TS, of which 33.5% was in suspended form, i.e., as TSS (Table S3 in SI). Organic matter content (expressed as COD) was high in influent samples (950 mg L^{-1}). Comparison of influent and effluent concentrations indicated that Assamra WWTP was efficient in TSS and COD removal (98% and 96%). At all sampling locations, along with ZR, the TSS concentrations were low (<2–40 mg L^{-1}) (Table S3), while the TS concentrations were high (840–4600 mg L^{-1}). Jerash stream contained the highest TS concentration (4600 mg L^{-1}) and had visible white residue. Organic matter content in river water was low (4–30 mg L^{-1}) at all sampling locations except Sukhna station, which seemed to have minor contamination with organic matter (110 mg COD L^{-1}). A wastewater pumping station is located upstream of Sukhna station, and leakages of wastewater would flow downstream to Sukhna station.

3.2. Concentration and Removal of PhACs and Antibiotics in Assamra WWTP

A total of 15 PhACs (excluding antibiotics) and one stimulant were detected in influent and effluent of Assamra WWTP. The combined concentration of PhACs (excluding antibiotics and caffeine; ∑PhACs) was 20,668–31,485 ng L^{-1} in the influent and 4032–4394 ng L^{-1} in the effluent, showing a significant reduction effect of treatment in Assamra WWTP (Figure 2, Table S4 in SI). The influent contained high levels of caffeine (27,737–53,223 ng L^{-1}), but these were reduced effectively in Assamra WWTP, resulting in a concentration of 64–273 ng L^{-1} in the effluent. Caffeine is highly biodegradable and can be used as an indicator of residual bioactivity [40].

The dominant PhACs in influent were paracetamol (anti-inflammatory; 14,891–24,309 ng L^{-1}) which comprised 74% of ∑PhACs, carbamazepine (anti-epileptic; 2365–3020 ng L^{-1}), which comprised 11% of ∑PhACs, atenolol (beta-blocker; 1723–1952 ng L^{-1}), which comprised 7% of ∑PhACs, and metformin (anti-diabetic), which comprised 4% of ∑PhACs. In the effluent, carbamazepine was the dominant PhAC (3138–3352 ng L^{-1}), comprising 78% of ∑PhACs, followed by metoprolol (beta-blocker; 10% of ∑PhACs). The removal rate of the dominant substances during WWTP treatment was: 99% for paracetamol, −22% for carbamazepine, 95% for atenolol, and 97% metformin (Figure 3). Poor removal has been reported previously for carbamazepine and hydrochlorothiazide in WWTPs in Spain [41]. Lajeunesse et al. also reported poor removal of carbamazepine in WWTPs in Canada [42]. Other studies investigating the removal of PhACs in middle-income countries (e.g., Jordan) report low removal efficiencies (<50%) for a number of PhACs, including carbamazepine [30]. Bisoprolol (beta-blocker) was present in higher effluent concentrations than influent (10-fold increase). Beta-blocker i.e bisoprolol is generally difficult to remove from wastewater [43]. The overall removal rate in Assamra WWTP was 81%–87% for ∑PhACs and 99% for caffeine. The concentrations of all measured substances are shown in Table S4.

In a previous study, Al-Mashaqbeh et al. [30] found that carbamazepine concentration was high in Assamra WWTP influent (1100 ng L^{-1}) and effluent (850 ng L^{-1}), resulting in low removal of the substance (23%). That study also reported a high occurrence of caffeine (156,000 ng L^{-1}) and its metabolite (1,7-dimethylxanthine; 10,500 ng L^{-1}) in influent, but high removal in Assamra WWTP plant (>99%), and very high concentrations (36,700 ng L^{-1}) of paracetamol in influent, but efficient removal in the WWTP (99%). The results in the present study confirm these findings of Al-Mashaqbeh et al. [30].

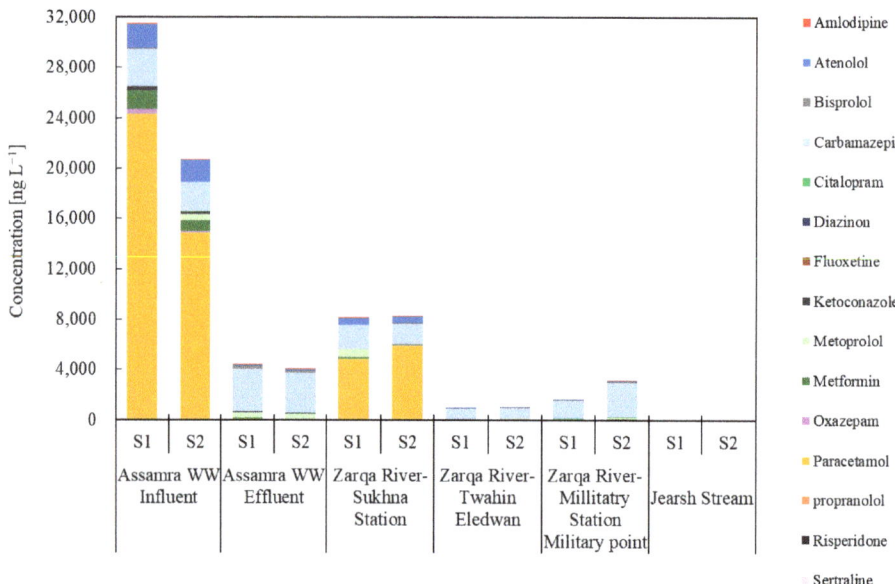

Figure 2. Concentrations (ng L^{-1}) of pharmaceutically active compounds (PhACs, excluding antibiotics) in Assamra wastewater treatment plant (WWTP) influent ($n = 2$) and effluent ($n = 2$), and in water from the Zarqa River at Sukhna station, ($n = 2$), Twahin Eledwan station ($n = 2$), Military station ($n = 2$) and Jerash stream ($n = 2$). S1, S2 = parallel samples.

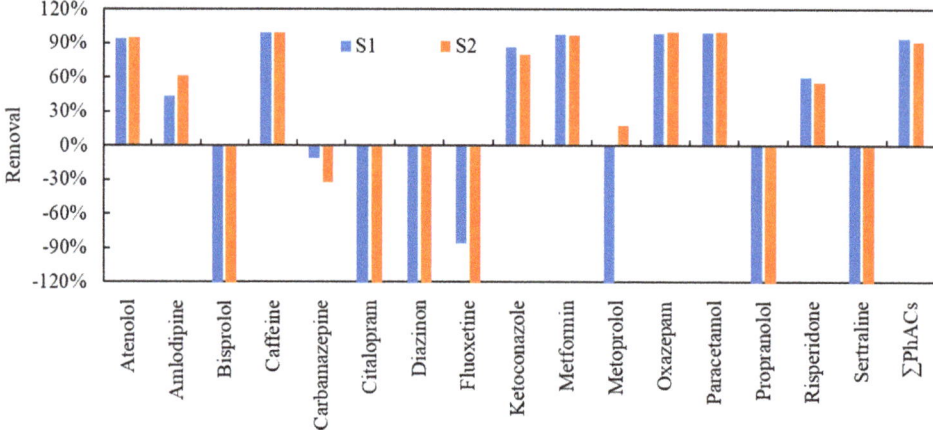

Figure 3. Efficiency (%) of Assamra wastewater treatment plant in removing pharmaceutically active compounds (PhACs, excluding antibiotics) from the wastewater. S1, S2 = parallel samples.

Of the 18 antibiotics targeted in the analysis, only five substances were detected in the influent and six in the effluent of Assamra WWTP. These were clarithromycin, erythromycin, ofloxacin, metronidazole, and sulfamethoxazole in the influent, and these five plus ciprofloxacin in the effluent (Figure 4). Ofloxacin (fluoroquinolones class) showed the highest concentration in wastewater effluent, followed by erythromycin (macrolides class) and metronidazole (antiprotozoal class) (Figure 4, Table S5 in SI). None of the target substances in the penicillin class (amoxicillin, ampicillin, benzylpenicillin) was detected in the influent or effluent samples. This result is surprising since Almaaytah et al. [2] reported that antibiotics in the classes fluoroquinolones, macrolides, penicillin, and antiprotozoal are the most commonly dispensed antibiotics in Jordan.

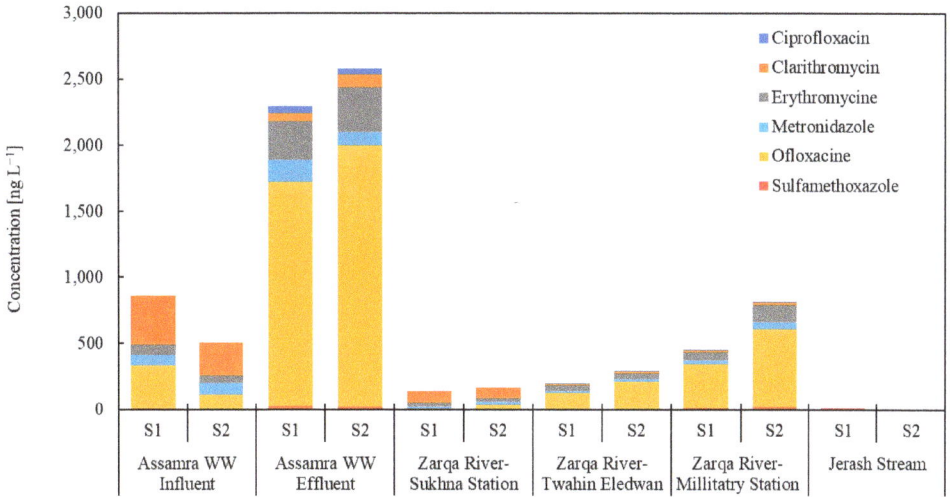

Figure 4. Concentrations of antibiotics (ng L^{-1}) in Assamra wastewater treatment plant (WWTP) influent (n = 2) and effluent (n = 2), and in water from the Zarqa River at Sukhna station, (n = 2), Twahin Eledwan station (n = 2), Military station (n = 2), and Jerash stream (n = 2). S1, S2 = parallel samples.

The combined antibiotics ($\sum$Antibiotics) concentration was 510–860 ng L^{-1} in influent and 2300–2600 ng L^{-1} in the effluent (Figure 4, Table S5). After passing through the treatment process, only the concentration of clarithromycin decreased (from 308 to 82 ng L^{-1}; removal rate 73% on average). C.F. Couto et al. reviewed the WWTPs performances to remove the PhACs in various countries [44]. The reported studies of 12 kinds of antibiotics, including Ofloxacin and Erythromycin, all observed positive values as removal ratio. However, the present study showed 2- to 5-fold higher concentrations of all other antibiotics in effluent than the levels in influent. Ciprofloxacin, which was not detected in the influent, was found in the effluent samples. Hydrolyses of organic matter of wastewater of solid phase and hence the release of ciprofloxacin bound to solids into the liquid phase could have occurred, explaining the increase of the substance in the effluent. Moreover, antibiotics partition into water is generally based on the chemical and physical properties of the antibiotic itself [45]. For example, clarithromycin has high partition coefficients (log Kow = 3.16, Koc = 150, Table S2) and low solubility. These properties suggest that clarithromycin is likely to adsorb to the solids in wastewater, which explains this substance's reduction. In contrast, ofloxacin has low log Kow (-2) and high-water solubility (6.7 × 10^5 mg L^{-1}) (Table S2), suggesting that it is water-soluble, explaining the high concentrations in the effluent.

In this study, we did not analyse the concentrations of antibiotics in the solid phase of raw wastewater or the WWTP sludge. It is likely that a fraction of the antibiotics initially partitioned to organic matter or accumulated in biomass in the wastewater was released after hydrolysis of the organic matter during biological treatment in the WWTP. This condition would partly explain the increase in antibiotic concentrations in effluent compared with influent, as also suggested in other studies [45,46]. Another explanation could be the presence of antibiotic conjugates or metabolites, which were cleaved back to their mother forms during the treatment process [46,47].

3.3. Spatial Distribution of PhACs and Antibiotics in ZR

At Sukhna station, on a tributary not impacted by Assamra WWTP effluent, $\sum$PhACs were 8141–8178 ng L^{-1} (Figure 2, Table S4). The high concentration of PhACs at this site was attributed to wastewater contamination and possible leakages of untreated wastewater from a pumping station upstream of Sukhna station. In addition, a small peri-urban community discharges its wastewater to the river, and sewage tankers illegally dump sewage close to the river. These are potential reasons for the high $\sum$PhACs in river water at Sukhna station.

The Twahin Eledwan station sampling site downstream of Assamra WWTP showed lower levels of $\sum$PhACs (970–1033 ng L^{-1}) than in WWTP effluent (4032–4394 ng L^{-1}). The dilution of PhACs can explain this, with surface runoff in the river channel. Thereafter, the concentration of $\sum$PhACs increased along the flow path in ZR and reached its highest level at Military station (1657–3154 ng L^{-1}) (Figure 1). Thus, there are evidently notable side-inputs of PhACs from areas surrounding the river. In particular, a military facility located downstream seems to be a point source of PhACs to ZR (Figures 3 and 4). In Jerash stream, a small downstream tributary of ZR, surface water samples showed low detectable deficient levels of $\sum$PhACs (<3 ng L^{-1}). The source of this tributary is spring water, and our sampling site was located just before the spring water flow mixed with other water from ZR.

It should be mentioned that the concentrations of individual PhACs differed significantly between all sampling sites. Paracetamol was detected at all sites but occurred in the highest concentrations at Sukhna station (4870–5900 ng L^{-1}). Paracetamol concentration then decreased to reach 40–70 ng L^{-1} as the water flowed downstream in ZR due to dilution and probable degradation. Carbamazepine was the dominant PhAC at all sites upstream and along ZR (800–2700 ng L^{-1}). Interestingly, we found a negative correlation between the concentration of carbamazepine and TSS content in water samples, i.e., the concentration of carbamazepine in water decreased as the TSS content in the water increased. Adsorption of carbamazepine to river sediment has been reported elsewhere [48,49]. Metoprolol and bisoprolol were among the dominant PhACs detected in ZR water.

The six antibiotic substances detected in Assamra WWTP effluent (clarithromycin, erythromycin, ofloxacin, metronidazole, sulfamethoxazole, and ciprofloxacin) were also detected in water at the different sampling sites along with ZR (Figure 4). Ofloxacin showed the highest concentrations in all sites along with ZR, and its concentration increased downstream in the river to reach the highest level at Military station (334–595 ng L^{-1}). It is not clear why this was the case. Still, we cannot exclude desorption of previously sorbed antibiotics from river sediment to water and illegal dumping of sewage sludge as contributing causes. Other studies have found that antibiotics (specifically sulfamethoxazole) decompose during transport within water systems [50]. In addition, the present study has shown the results of one-time sampling, but the river flow rate is subjected to seasonal change that would affect the concentration and retention time of the PhACs [51]. Therefore, more monitoring campaigns are recommended for future work.

The estimated mass flow of target substances in ZR water at Military station was 39–71 kg year^{-1} for antibiotics and 60–110 kg year^{-1} for other PhACs, based on water flow of 86×10^6 m^3 year^{-1}. It should be pointed out that most of the water in ZR is used for the irrigation of vegetables and fodder crops [35]. The fate of antibiotics and other PhACs in irrigated soils in the study area was not analysed in this study. However, the transport of antibiotics within the ZR water system is alarming and there is likelihood that it poses a risk of developing and spreading antimicrobial resistance within the area. Upstream measures might be needed to reduce antibiotics in wastewater and limit the loads entering the water system and the environment. Such measures could include limiting the prescription and sale of antibiotics and increasing awareness among the public and pharmacists of the consequences of antimicrobial resistance.

4. Conclusions

This study examined the occurrence of 15 pharmaceutically active compounds (PhACs), 18 antibiotics, and one stimulant (caffeine) in influent and effluent from Assamra WWTP and in water samples from the Zarqa River and its tributaries. Based on the results obtained, the following conclusions were drawn:

- Assamra WWTP is efficient in minimising the concentrations of PhACs and caffeine, with an overall efficiency of 81–87% and 99%, respectively. It is inefficient in removing bisoprolol and carbamazepine, for which effluent concentrations were 128% and 22% higher than influent.
- Assamra WWTP is inefficient in removing antibiotics from wastewater, as concentrations of all antibiotics detected, except clarithromycin, increased by 2- to 5-fold in the effluent compared with the influent.
- Zarqa river water is contaminated with antibiotics and PhACs. Sources of these contaminants are likely to be effluent from Assamra WWTP and side-inputs from the areas surrounding the river.
- $\sum$Antibiotics and $\sum$PhACs in Zarqa river water are still lower than those in Assamra WWTP effluent. Dilution, degradation, evaporation, and adsorption are potential mechanisms contributing to lowering the concentration of PhACs along the river.
- Since most Zarqa river is used for irrigation of vegetables and fodder crops, the PhACs and antibiotics in river water could enter the food chain and pose a risk of spreading antibiotic-resistant genes and mobile genetic elements (i.e., plasmids and integrin). Further research is required to study the fate of antibiotics and other PhACs in irrigated soils, WWTP sludge, and plants in the Zarqa river basin.
- Future research should aim at conducting more sampling and at staging analysis at different seasons to understand the effects of weather conditions on the fate of PhACs in the study area. In addition, analyses of the PhACs in the solid phase (i.e., sludge) are necessary to understand the partitioning of the PhACs between the water and solid phases of the wastewater in the Assamra plant and to evaluate the total loads of PhACs in the wastewater influent and effluent.

Supplementary Materials: The following are available online at https://www.mdpi.com/2076-3417/11/18/8638/s1, Table S1. Geographic positioning system (GPS) coordinates of the sampling sites for wastewater treatment plant (WWTP) influent and effluent and Zarqa River (ZR) water. Data from [35], Table S2. List of pharmaceutically active compounds (PhACs, including antibiotics) targeted in analysis of liquid samples and their chemical and physical properties (chemical formula, molecular weight (MW), logarithmic octanol-water distribution coefficient (log Kow), organic carbon-water partition coefficient (Koc), logarithmic dissociation constant (pKa), and water solubility at 25 °C (mg/L). Values without any reference superscript (+) were modeled and taken from ChemSpider (2020), Table S3. Concentrations of total suspended solids (TSS), total solids (TS) and chemical oxygen demand (COD) samples (n = 1 per site) in Assamra wastewater treatment plant (WWTP) influent and effluent and Zarqa River (ZR) water at Sukhna station, Twahin Eledwan station, Military station, and Jerash stream. Data from [35].

Author Contributions: The authors had contributed to the published work according to the following description: M.S.: conducted the laboratory analyses, the statistical analyses and revised the manuscript according to reviewers input; A.A.: participated in design and planning of the study; A.H.: participated in describing and visualizing the study area, S.S.D.: planned the study, collected the samples, participated in data analyses and wrote the manuscript. All authors have read and agreed to the published version of the manuscript.

Funding: The study was funded by the Swedish Research Council (FORMAS) through grant 169-2013-1963 for the project "Pharmaceutical Pollution at Use of Wastewater in Crop Production" and the Swedish Foundation for international cooperation in research and education (grant number PT2016–6875).

Data Availability Statement: Original data are presented in the supporting information.

Acknowledgments: Special thanks to Raed Jaber, Samir Tawalbeh, and Mohamed Mashatleh at the Royal Scientific Society of Jordan for their assistance in sample collection and facilitating the shipping of samples to Sweden. Thanks also to members of the Water Quality Laboratory at the Royal Scientific Society of Jordan for their efforts in physical and chemical analyses of the samples. Instrumental identification of the pharmaceuticals was performed at the Swedish Environmental Institute (IVL), Stockholm. Special thanks to Gunnar Thorsén for his effort in the analytical procedures.

Conflicts of Interest: The authors have no competing interest to declare, or financial or personal relationships with other people or organizations that could have inappropriately influenced this work.

References

1. Al-Azzam, S.I.; Al-Husein, B.A.; Alzoubi, F.; Masadeh, M.M.; Al-Horani, M.A.S. Self-medication with antibiotics in Jordanian population. *Int. J. Occup. Med. Environ. Health* **2007**, *20*, 373–380. [CrossRef]
2. Almaaytah, A.; Mukattash, T.L.; Hajaj, J. Dispensing of non-prescribed antibiotics in Jordan. *Patient Prefer. Adherence* **2015**, *9*, 1389–1395. [CrossRef]
3. Nimri, L.F.; Batchoun, R. Community-acquired bacteraemia in a rural area: Predominant bacterial species and antibiotic resistance. *J. Med. Microbiol.* **2004**, *53*, 1045–1049. [CrossRef]
4. Shakhatreh, M.; Swedan, S.; Al-Odat, M.; Khabour, O. Uropathogenic *Escherichia coli* (UPEC) in Jordan: Prevalence of Urovirulence Genes and Antibiotic Resistance. *J. King Saud Univ. Sci.* **2018**, *31*, 648–652. [CrossRef]
5. Bijlsma, L.; Emke, E.; Hernández, F.; de Voogt, P. Investigation of drugs of abuse and relevant metabolites in Dutch sewage water by liquid chromatography coupled to high resolution mass spectrometry. *Chemosphere* **2012**, *89*, 1399–1406. [CrossRef]
6. Lindberg, R.; Östman, M.; Olofsson, U.; Grabic, R.; Fick, J. Occurrence and behaviour of 105 active pharmaceutical ingredients in sewage waters of a municipal sewer collection system. *Water Res.* **2014**, *58*, 221–229. [CrossRef]
7. D'Sa, S.; Patnaik, D. The Impact of the Pharmaceutical Industry of Hyderabad in the Pollution of the Godavari River. In *Water Management in South Asia: Socio-Economic, Infrastructural, Environmental and Institutional Aspects*; Bandyopadhyay, S., Magsi, H., Sen, S., Ponce Dentinho, T., Eds.; Contemporary South Asian Studies; Springer: Cham, Switzerland, 2020; pp. 23–51. ISBN 978-3-030-35237-0.
8. Kabdaşlı, N.; Olmez-Hanci, T.; Akgun, G.; Tunay, O. Assessment of Pollution Profile and Wastewater Control Alternatives of a Pharmaceutical Industry. *Fresenius Environ. Bull.* **2019**, *28*, 626–632.
9. Velpandian, T.; Halder, N.; Nath, M.; Das, U.; Moksha, L.; Gowtham, L.; Batta, S.P. Un-segregated waste disposal: An alarming threat of antimicrobials in surface and ground water sources in Delhi. *Environ. Sci. Pollut. Res. Int.* **2018**, *25*, 29518–29528. [CrossRef]
10. Garrido-Cardenas, J.A.; Esteban-García, B.; Agüera, A.; Sánchez-Pérez, J.A.; Manzano-Agugliaro, F. Wastewater Treatment by Advanced Oxidation Process and Their Worldwide Research Trends. *Int. J. Environ. Res. Public. Health* **2020**, *17*, 170. [CrossRef] [PubMed]
11. Ankush; Mandal, M.K.; Sharma, M.; Khushboo; Pandey, S.; Dubey, K.K. Membrane Technologies for the Treatment of Pharmaceutical Industry Wastewater. In *Water and Wastewater Treatment Technologies*; Bui, X.-T., Chiemchaisri, C., Fujioka, T., Varjani, S., Eds.; Energy, Environment, and Sustainability; Springer: Singapore, 2019; pp. 103–116. ISBN 9789811332593.
12. Rivadulla, E.; García-Fernández, I.; Romalde, J.; Fernandez-Ibanez, P.; Polo, D. Solar radiation and photo-Fenton systems: Novel applications for norovirus disinfection in water. In *Noroviruses: Outbreaks, Control and Prevention Strategies*; Nova Science Publishers: Hauppauge, NY, USA, 2017.
13. Sudeep, M.; Vinutha, M. Nanotechnology-As antibacterial and heavy metal removal in waste water treatment-A review. *AIP Conf. Proc.* **2018**, *2039*, 020067. [CrossRef]
14. Bougnom, B.P.; Zongo, C.; McNally, A.; Ricci, V.; Etoa, F.X.; Thiele-Bruhn, S.; Piddock, L.J.V. Wastewater used for urban agriculture in West Africa as a reservoir for antibacterial resistance dissemination. *Environ. Res.* **2019**, *168*, 14–24. [CrossRef]
15. Graham, D.W.; Giesen, M.J.; Bunce, J.T. Strategic Approach for Prioritising Local and Regional Sanitation Interventions for Reducing Global Antibiotic Resistance. *Water* **2019**, *11*, 27. [CrossRef]
16. Marathe, N.P.; Pal, C.; Gaikwad, S.S.; Jonsson, V.; Kristiansson, E.; Larsson, D.G.J. Untreated urban waste contaminates Indian river sediments with resistance genes to last resort antibiotics. *Water Res.* **2017**, *124*, 388–397. [CrossRef]
17. Reymond, P.; Abdel Wahaab, R.; Moussa, M.S.; Lüthi, C. Scaling up small scale wastewater treatment systems in low- and middle-income countries: An analysis of challenges and ways forward through the case of Egypt. *Util. Policy* **2018**, *52*, 13–21. [CrossRef]
18. Arya, G.; Tadayon, S.; Sadighian, J.; Jones, J.; de Mutsert, K.; Huff, T.B.; Foster, G.D. Pharmaceutical chemicals, steroids and xenoestrogens in water, sediments and fish from the tidal freshwater Potomac River (Virginia, USA). *J. Environ. Sci. Health Part A Toxic Hazard. Subst. Environ. Eng.* **2017**, *52*, 686–696. [CrossRef]
19. Burke, V.; Richter, D.; Greskowiak, J.; Mehrtens, A.; Schulz, L.; Massmann, G. Occurrence of Antibiotics in Surface and Groundwater of a Drinking Water Catchment Area in Germany. *Water Environ. Res. Res. Publ. Water Environ. Fed.* **2016**, *88*, 652–659. [CrossRef]

20. Dodgen, L.K.; Kelly, W.R.; Panno, S.V.; Taylor, S.J.; Armstrong, D.L.; Wiles, K.N.; Zhang, Y.; Zheng, W. Characterizing pharmaceutical, personal care product, and hormone contamination in a karst aquifer of southwestern Illinois, USA, using water quality and stream flow parameters. *Sci. Total Environ.* **2017**, *578*, 281–289. [CrossRef]
21. Jurado, A.; Walther, M.; Díaz-Cruz, M.S. Occurrence, fate and environmental risk assessment of the organic microcontaminants included in the Watch Lists set by EU Decisions 2015/495 and 2018/840 in the groundwater of Spain. *Sci. Total Environ.* **2019**, *663*, 285–296. [CrossRef] [PubMed]
22. König, M.; Escher, B.I.; Neale, P.A.; Krauss, M.; Hilscherová, K.; Novák, J.; Teodorović, I.; Schulze, T.; Seidensticker, S.; Kamal Hashmi, M.A.; et al. Impact of untreated wastewater on a major European river evaluated with a combination of in vitro bioassays and chemical analysis. *Environ. Pollut.* **2017**, *220*, 1220–1230. [CrossRef]
23. Nishi, I.; Kawakami, T.; Onodera, S. Monitoring the concentrations of nonsteroidal anti-inflammatory drugs and cyclooxygenase-inhibiting activities in the surface waters of the Tone Canal and Edo River Basin. *J. Environ. Sci. Health Part A Toxic. Hazard. Subst. Environ. Eng.* **2015**, *50*, 1108–1115. [CrossRef]
24. Schimmelpfennig, S.; Kirillin, G.; Engelhardt, C.; Duennbier, U.; Nützmann, G. Fate of pharmaceutical micro-pollutants in Lake Tegel (Berlin, Germany): The impact of lake-specific mechanisms. *Environ. Earth Sci.* **2016**, *75*, 893. [CrossRef]
25. Directorate of Media and Water Awareness. *The Annual Report of Ministry of Water and Irrigation*; Ministry of Water and Irrigation (MWI): Amman, Jordan, 2018.
26. Al-Omari, A.; Farhan, I.; Kandakji, T.; Jibril, F. Pollution Sources to Zarqa River: Their Impact on the River Water Quality as a Source of Irrigation Water. 2017. Available online: http://mena.exceed-swindon.org/wp-content/uploads/2015/03/0025-Abbas-Manuscript-Marakech-2018-02-27-final-pp.pdf (accessed on 3 September 2021).
27. Tiehm, A.; Schmidt, N.; Lipp, P.; Zawadsky, C.; Marei, A.; Seder, N.; Ghanem, M.; Paris, S.; Zemann, M.; Wolf, L. Consideration of emerging pollutants in groundwater-based reuse concepts. *Water Sci. Technol.* **2012**, *66*, 1270–1276. [CrossRef]
28. Tiehm, A.; Schmidt, N.; Stieber, M.; Sacher, F.; Wolf, L.; Hoetzl, H. Biodegradation of Pharmaceutical Compounds and their Occurrence in the Jordan Valley. *Water Resour. Manag.* **2011**, *25*, 1195–1203. [CrossRef]
29. Zemann, M.; Wolf, L.; Pöschko, A.; Schmidt, N.; Sawarieh, A.; Seder, N.; Tiehm, A.; Hötzl, H.; Goldscheider, N. Sources and processes affecting the spatio-temporal distribution of pharmaceuticals and X-ray contrast media in the water resources of the Lower Jordan Valley, Jordan. *Sci. Total Environ.* **2014**, *488–489*, 100–114. [CrossRef]
30. Al-Mashaqbeh, O.; Alsafadi, D.; Dalahmeh, S.; Bartelt-Hunt, S.; Snow, D. Correction: Almashaqbeh, O.; et al., Removal of Selected Pharmaceuticals and Personal Care Products in Wastewater Treatment Plant in Jordan. *Water* **2020**, *12*, 1122. [CrossRef]
31. Al-Omari, A.; Farhan, I.; Kandakji, T.; Jibril, F. Zarqa River pollution: Impact on its quality. *Environ. Monit. Assess.* **2019**, *191*, 166. [CrossRef]
32. *Adaptation to Climate Change to Sustain Jordan's MDG Achievements*; Ministry of Water and Irrigation (MWI): Amman, Jordan, 2013.
33. SUEZ. *As Samra Wastewater Treatment Plant (Jordan)*; SUEZ: Amman, Jordan, 2019; p. 7.
34. Myszograj, S.; Qteishat, O. Operate of As-Samra Wastewater Treatment Plant in Jordan and Suitability for Water Reuse. *Inż. Ochr. Śr.* **2011**, *14*, 29–40.
35. Shigei, M.; Ahrens, L.; Hazaymeh, A.; Dalahmeh, S.S. Per- and polyfluoroalkyl substances in water and soil in wastewater-irrigated farmland in Jordan. *Sci. Total Environ.* **2020**, *716*, 137057. [CrossRef]
36. Dalahmeh, S.; Ahrens, L.; Gros, M.; Wiberg, K.; Pell, M. Potential of biochar filters for onsite sewage treatment: Adsorption and biological degradation of pharmaceuticals in laboratory filters with active, inactive and no biofilm. *Sci. Total Environ.* **2018**, *612*, 192–201. [CrossRef]
37. Dalahmeh, S.S.; Assayed, A.; Stenström, Y. Combined Vertical-Horizontal Flow Biochar Filter for Onsite Wastewater Treatment—Removal of Organic Matter, Nitrogen and Pathogens. *Appl. Sci.* **2019**, *9*, 5386. [CrossRef]
38. Gros, M.; Ahrens, L.; Levén, L.; Koch, A.; Dalahmeh, S.; Ljung, E.; Lundin, G.; Jönsson, H.; Eveborn, D.; Wiberg, K. Pharmaceuticals in source separated sanitation systems: Fecal sludge and blackwater treatment. *Sci. Total Environ.* **2019**, *703*, 135530. [CrossRef]
39. Eaton, A.D.; Clesceri, L.S.; Greenberg, A.E.; Franson, M.A.H. American Public Health Association; American Water Works Association; Water Environment Federation. *Standard Methods for the Examination of Water and Wastewater*; American Public Health Association: Washington, DC, USA, 1998; ISBN 978-0-87553-235-6.
40. Buerge, I.I.; Poiger, T.; Müller, M.D.; Buser, H.-R. Caffeine, an anthropogenic marker for wastewater comtamination of surface waters. *Environ. Sci. Technol.* **2003**, *37*, 691–700. [CrossRef]
41. Gros, M.; Rodríguez-Mozaz, S.; Barceló, D. Rapid analysis of multiclass antibiotic residues and some of their metabolites in hospital, urban wastewater and river water by ultra-high-performance liquid chromatography coupled to quadrupole-linear ion trap tandem mass spectrometry. *J. Chromatogr. A* **2013**, *1292*, 173–188. [CrossRef]
42. Lajeunesse, A.; Smyth, S.A.; Barclay, K.; Sauvé, S.; Gagnon, C. Distribution of antidepressant residues in wastewater and biosolids following different treatment processes by municipal wastewater treatment plants in Canada. *Water Res.* **2012**, *46*, 5600–5612. [CrossRef]
43. Baresel, C.; Cousins, A.P.; Hörsing, M.; Ek, M.; Ejhed, H.; Allard, A.-S.; Magnér, J.; Westling, K.; Wahlberg, C.; Fortkamp, U.; et al. *Pharmaceutical Residues and Other Emerging Substances in the Effluent of Sewage Treatment Plants*; Swedish Environmental Reserch Institute: Stockholm, Sweden, 2015; p. 118.
44. Couto, C.F.; Lange, L.C.; Amaral, M.C.S. Occurrence, fate and removal of pharmaceutically active compounds (PhACs) in water and wastewater treatment plants—A review. *J. Water Process Eng.* **2019**, *32*, 100927. [CrossRef]

45. Kulkarni, P.; Olson, N.D.; Raspanti, G.A.; Rosenberg Goldstein, R.E.; Gibbs, S.G.; Sapkota, A.; Sapkota, A.R. Antibiotic Concentrations Decrease during Wastewater Treatment but Persist at Low Levels in Reclaimed Water. *Int. J. Environ. Res. Public. Health* **2017**, *14*, 668. [CrossRef] [PubMed]
46. Le-Minh, N.; Khan, S.J.; Drewes, J.E.; Stuetz, R.M. Fate of antibiotics during municipal water recycling treatment processes. *Water Res.* **2010**, *44*, 4295–4323. [CrossRef] [PubMed]
47. Al-Tarawneh, I.; El-Dosoky, M.; Alawi, M.; Batarseh, M.; Widyasari, A.; Kreuzig, R.; Bahadir, M. Studies on Human Pharmaceuticals in Jordanian Wastewater Samples. *CLEAN–Soil Air Water* **2015**, *43*, 504–511. [CrossRef]
48. Carmona, E.; Andreu, V.; Picó, Y. Occurrence of acidic pharmaceuticals and personal care products in Turia River Basin: From waste to drinking water. *Sci. Total Environ.* **2014**, *484*, 53–63. [CrossRef] [PubMed]
49. Yang, Y.-Y.; Toor, G.; Williams, C. Pharmaceuticals and organochlorine pesticides in sediments of an urban river in Florida, USA. *J. Soils Sediments* **2015**, *15*, 993–1004. [CrossRef]
50. Kadlec, R.H.; Wallace, S.D. *Treatment Wetlands*, 2nd ed.; CRC Press: Boca Raton, FL, USA, 2009; ISBN 978-1-56670-526-4.
51. Mandaric, L.; Kalogianni, E.; Skoulikidis, N.; Petrovic, M.; Sabater, S. Contamination patterns and attenuation of pharmaceuticals in a temporary Mediterranean river. *Sci. Total Environ.* **2019**, *647*, 561–569. [CrossRef] [PubMed]

Article

Long-Term Occurrence and Fate of Microplastics in WWTPs: A Case Study in Southwest Europe

Andrea Menéndez-Manjón [1], Reyes Martínez-Díez [1], Daniel Sol [1], Amanda Laca [1], Adriana Laca [1,*], Amador Rancaño [2] and Mario Díaz [1]

[1] Department of Chemical and Environmental Engineering, University of Oviedo, C/Julián Clavería s/n, 33006 Oviedo, Spain; andreamanjon95@gmail.com (A.M.-M.); reyes.m.d@hotmail.com (R.M.-D.); dsolsan91@gmail.com (D.S.); lacaamanda@uniovi.es (A.L.); mariodiaz@uniovi.es (M.D.)
[2] ACCIONA Agua S.A., 28108 Alcobendas, Spain; amador.rancano.perez@acciona.com
* Correspondence: lacaadriana@uniovi.es; Tel.: +34-985-10-29-74

Abstract: Microplastic (MP) water pollution is a major problem that the world is currently facing, and wastewater treatment plants (WWTPs) represent one of the main alternatives to reduce the MP release to the environment. Several studies have analysed punctual samples taken throughout the wastewater treatment line. However, there are few long-term studies on the evolution of MPs over time in WWTPs. This work analyses the performance of a WWTP sited in Southwest Europe in relation with annual occurrence and fate of MPs. Samples were monthly taken at different points of the facility (influent, secondary effluent, final effluent, and sludge) and MPs were quantified and characterised by means of stereomicroscopy and FTIR spectrophotometry. The majority of MPs found in wastewater and sludge samples were fragments and fibres. Regarding to the chemical composition, in the water samples, polyethylene (PE), polyethylene terephthalate (PET) and polypropylene (PP) stood out, whereas, in the sludge samples, the main polymers were PET, polyamide (PA) and polystyrene (PS). The MPs more easily removed during the wastewater treatment processes were those with sizes greater than 500 μm. Results showed that the MPs removal was very high during all the period analysed with removal efficiencies between 89% and 95%, so no great variations were found between months. MP concentrations in dry sludge samples ranged between 12 and 39 MPs/g, which represented around 79% of the total MPs removed during the wastewater treatment processes. It is noticeable that a trend between temperature and MPs entrapped in sewage sludge was observed, i.e., higher temperatures entailed higher percentage of retention.

Keywords: microplastics; sludge; WWTP; removal efficiency; secondary treatment

1. Introduction

WWTPs are a major indirect source of MP emissions into the environment, due to the daily discharge of large quantities of MPs, from agricultural, industrial or urban activities, to the sewage system [1–3]. At the household level, this pollution mainly comes from the use of products that containing MPs, namely cosmetic and personal care products, and also fibres generated during laundry [4–6]. In addition, MPs can be originated from the weathering and fragmentation of plastics due to disposal mismanagement or by the wear and tear of plastic items [7–9]. These microplastics can enter to the sewage system by surface runoff or stormwater, either because they are on the ground surface or deposited from the atmosphere [10–12]; therefore, wastewater could contain a high number of MPs, specifically, the MP concentration reported in WWTPs ranged between 0.28 and 3.14·10^4 particles/L [13]. Although WWTPs can frequently achieve removal efficiencies of MPs up to 90%, this is insufficient because large quantities of microplastics are still being released into rivers and oceans [13–15].

It has been reported that most MPs removed during the wastewater treatment are accumulated in sludge [16]. So far, the reported ranges of MP concentration in wet and dry mixed sludge were 400–7000 and 1500–170,000 particles/kg, respectively [17–20]. Furthermore, the repeated application of sludge in agriculture as soil amendment is a potential problem, as it favours the excessive and unavoidable accumulation of MPs in the farmlands. It is estimated that the use of sludge as fertilizer releases in European agricultural lands between 63,000 and 430,000 tons of MPs per year [21,22]. MPs not removed from the wastewater during the treatment processes are finally released into the aquatic environment; in particular, the abundance of MPs in the effluent of urban WWTPs ranges between 0.01 and 297 particles/L [13]. MPs emitted to the environment become a potential risk, not only to the ecosystems, but also to human health, since they can be bioaccumulated through the trophic chain [23–26].

Several chemical, physical and biological processes take place in WWTPs to achieve high-quality effluent water. Each treatment plant uses its own technologies depending on different factors (the subsequent reuse of water, the characteristics of wastewater, the place where the effluent is discharged, etc.) [3]. When the wastewater treatment includes dynamic membranes (DMs) or membrane bioreactor (MBR), MP removals of 99% or even higher have been reported [18,27–30]. The major drawback is the high cost of implementing and maintaining these technologies. Surprisingly, there are some works that reported similar removal efficiencies employing lower cost technologies, such as conventional activated sludge (CAS) and sequencing batch reactor (SBR) [18,31]. In fact, removal efficiencies in the range between 96–98% have been reported from WWTPs that used that kind of technologies. It is necessary to point out that most works have calculated the removal efficiencies just by analysing a few samples, which can contribute to the dispersion of efficiencies. Analysing the WWTPs performance for extended periods would be necessary to stablish accurate conclusions. Therefore, in this work, the performance of wastewater treatment processes was evaluated in a WWTP sited in Southwest Europe over a 12-month period. The aim of the study is increasing the knowledge on the behaviour, fate and elimination of microplastics in the different stages of treatment throughout the year. Furthermore, as far as we know, it is the first study to analyse the effect of a double consecutive decantation (secondary treatment), as well as the use of a lamellar settler in the tertiary treatment.

2. Materials and Methods

2.1. WWTP Characteristics

The WWTP is located in the Southwest of Spain, specifically in Caravaca de la Cruz (Murcia). It was designed to treat an average daily flow of 8000 m^3, serving 85,000 population equivalent (p.e.). Firstly, as can be seen in Figure 1, the raw water is pre-treated by means of a screening system (pore size of 10 mm and 3 mm) and an aerated grit and grease removal system. The secondary treatment consists of an anoxic tank with a capacity of 950 m^3 with two agitators, two carousel-type aeration tanks with a total volume of 19,000 m^3 and two secondary decanters placed in series. Finally, the tertiary treatment consists of coagulation-flocculation tank, lamellar decanter, rapid sand filter (RSF) and UV disinfection system.

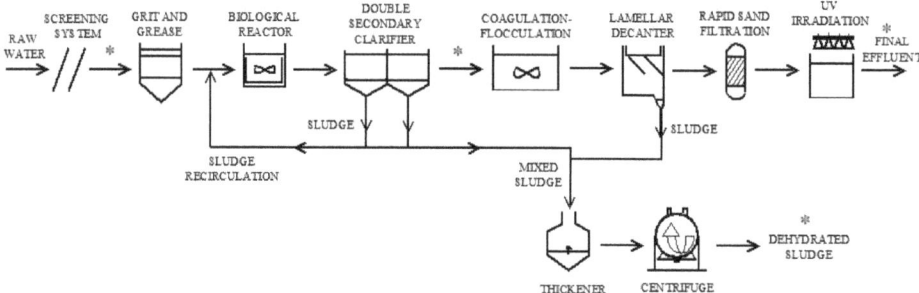

Figure 1. Scheme of WWTP analysed in this work (asterisks indicate sampling points).

The sludge recovered from the secondary and tertiary treatment is mixed, thickened by settling, and finally, dewatered by means of centrifugation, up to 78–86% moisture content (w/w).

2.2. Sampling Points

To obtain representative and homogeneous samples, the water was collected in turbulent areas to prevent the heavier MPs to sediment. Sampling points are indicated by an asterisk in Figure 1, i.e., after the screening systems (influent), effluent from secondary treatment, effluent from tertiary treatment and dehydrated sludge. To collect the water, it was pumped through a sieve module assembled in a specific sampling device (Figure S1). This device is made up of four mesh stainless steel filters (CISA Sieving Technologies) of 150 mm of diameter and the following slot sizes: 500, 250, 100 and 20 µm (placed from the largest to the smallest one). Thus, MPs contained in the sampled water were classified by size and retained in the corresponding sieve. The flow rate chosen for sampling was 10 L/min which was maintained during approximately 30 min (or until the solids clog the sieves) at each collection point. MPs collected were dragged with distilled water and stored under refrigeration until further processing. The volumes of wastewater sampled at each sampling point each month are detailed in Table S1.

Dewatered sludge samples were also stored under refrigeration. In order to express MP concentration on dry weight basis (w/w), a gravimetric method was used to determine the moisture content of each sample of sludge by triplicate.

2.3. Pre-Treatment of Samples

Water samples were stored in an oven at 90 °C to dryness. After that, the organic matter was degraded by treating the samples with Fenton's reagent (20 mL of solution of Fe(II) at pH 3 with 20 mL of H_2O_2 50%) at room temperature, during 30 min. Once digested, samples were left at room temperature for 24 h to allow the residual hydrogen peroxide to evaporate and, then, they were stored in an oven at 90 °C to dryness (10 h). MPs were isolated from the remaining inorganic impurities by density using a solution of $ZnCl_2$ (d = 1.6 g/mL) (97% purity, VWR), so that supernatant was filtered under vacuum using a glass microfiber filter (Whatman, diameter 47 mm, pore size of 0.7 µm).

Sludge samples (5 g) were oxidised during 24 h with 30 mL of hydrogen peroxide (H_2O_2, 50%). This process was carried out twice. The rest of the procedure was the same as that employed for water samples.

Distilled water and zinc chloride solution employed in the pre-treatment samples were previously filtered using a glass microfiber filter (Whatman, diameter 47 mm, pore size of 0.7 µm) to avoid MP contamination.

2.4. Microplastic Analysis

Filters with MPs were examined under a semiautomatic stereomicroscope (Leica M205FA) with a high-resolution colour digital camera attached (Leica DFC310FX) to process images with a maximum resolution of 1392 × 1040 pixels (1.4 Mpixels CCD). It is used for the quantification of MPs and the analysis of colour and shape of microparticles [28,32–34].

To determine the chemical composition of microplastics, an FTIR spectrophotometer coupled to a microscope with an imaging system (Varian 620-IR and Varian 670-IR) with three detection systems is used [35]. Samples were analysed in the mid-infrared of 4000–400 cm^{-1}, a range in which the most typical bands of plastic materials are identified. The identification of functional groups and molecular composition of polymeric surfaces was carried out using the list of absorption bands of sixteen polymers described by Jung et al. (2018) [36].

3. Results and Discussion
3.1. Occurrence and Evolution of MPs

Nowadays, most of the studies dealing with the occurrence of MPs in WWTPs have been focused on wastewater and sludge samples collected over short periods, i.e., days or weeks [16,32,37]. In this work, the occurrence and evolution of microplastics in a WWTP have been examined, over a 12-month period (Figure 2). Figure 2a summarised the MP concentration in the different sampling points in the WWTP analysed during the study.

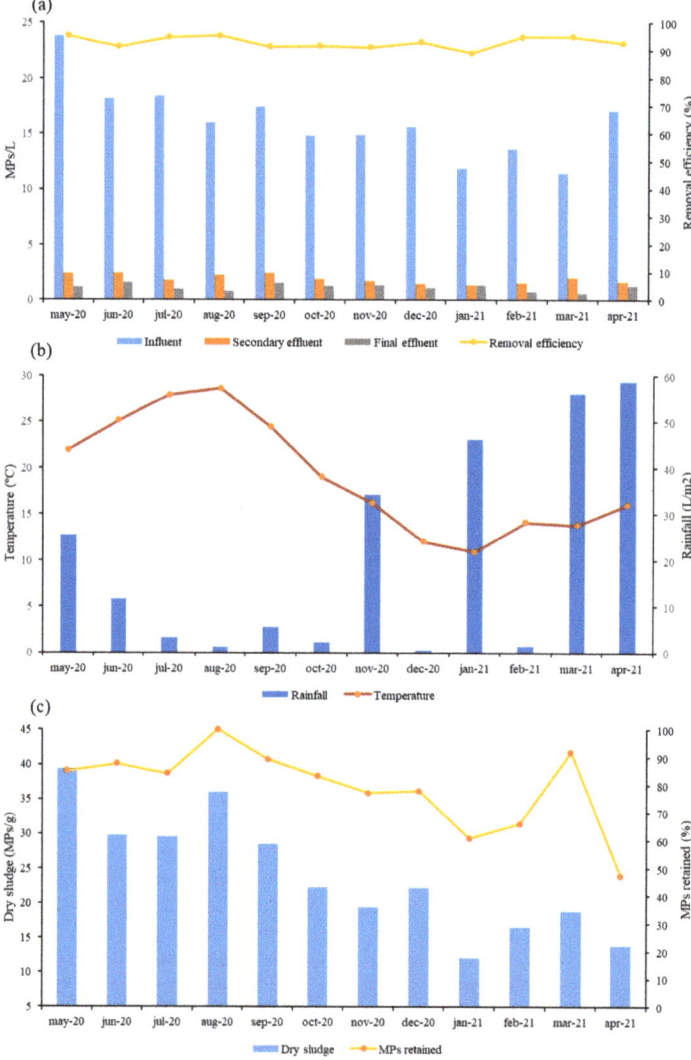

Figure 2. (a) Microplastic concentration (MPs/L) in the WWTP at the four sampling points and overall removal efficiency for one year (May 2020–April 2021), (b) rainfall (L/m^2) and temperature (°C) recorded in Caravaca de la Cruz over the sampling period (Source: State Meteorological Agency [38]), (c) Microplastic concentration in sludge expressed per dry weight (MPs/g) and percentages of MPs retained in the sludge with respect to the MPs removed during the treatment in the WWTP during the period studied.

The WWTP receives in the influent, after screening systems, a mean concentration of 16.1 ± 3.3 MPs/L. This value is in accordance with other studies reporting similar MP concentrations in the influent of urban WWTP, for example, between 12–16 MPs/L in China [39–41], 12.2 MPs/L in Thailand [42], 15.1 MPs/L in Sweden [43] and 15.7 MPs/L in Scotland [44]. Nevertheless, it should be noted that other works reported MP concentration in influent samples much higher [45] or slightly lower [46] than those found in this work. This can be since the number of MPs in wastewater can be affected by different factors such as population served, lifestyle, climate and seasonal conditions [47].

Considering Figure 2, it can be observed that during the warmest months, from April to September, the MP concentration in influents is, in general, slightly higher compared to the coldest, i.e., January to March. This is probably due to the higher evaporation of water that concentrates microplastics in the aqueous stream. This is in agreement with previous studies, carried out in Spain [10]. It may be due to the fragmentation of (micro)plastics by greater solar irradiation and, to the increase of MP concentration by evaporation of water. On the contrary, Ben-David et al. [48] studied a WWTP in a city located in the north of Israel that reported higher values of MP concentration in the rainy winter season, which was associated with a greater use of washing machines or a greater contribution from land runoff. In this case, there is not any clear correlation between rainfall and the MP concentrations found in influent.

After pre-treatment, the secondary treatment consists of a biological reactor together with a double settling tank. So far, there is no literature data reported on the effect of a double decanter for MP elimination. In general, secondary effluent shows a notable decrease in MP concentration in comparison with those in influent (an average value of 1.90 ± 0.38 MPs/L), which means a removal efficiency (grit and grease removal + biological treatment) higher than of 88%. Hidayaturrahman and Lee [49] analysed the influence of grit and grease and secondary treatment in three WWTPs with MP removals between 75–93%. Similar results were obtained by Ruan et al. [50], who found elimination efficiencies of 87%, whereas Yang et al. [41], after secondary treatment, obtained a removal efficiency of 72%.

It is clear that WWTPs with tertiary treatments have been reported to be more efficient in eliminating MPs than systems that present only a secondary treatment [51]. For example, Magni et al. [19] found a removal efficiency of 64% after the secondary treatment and 84% after the tertiary. In addition, Ziajahromi et al. [52] indicated that, after the secondary treatment, the removal efficiency of MPs was 66%, whereas, after a tertiary treatment, was 87%. Regarding the tertiary treatment applied in this WWTP, that consists of a coagulation-flocculation, a lamellar settler, a RSF and an UV disinfection, the removal efficiency of MPs of around 41% was achieved, which increased the overall removal efficiency until values of around 93% and entails an emission of 1.13 MPs/L in the effluent. In those effluent samples, during the warmest months (April to September) the MP concentration was higher compared to the coldest ones (January to March) with ranges of 0.77–1.58 MPs/L (1.21 ± 0.31) and 0.59–1.31 MPs/L (0.87 ± 0.38), respectively. These results are in accordance with those reported by Jiang et al. [53].

Although coagulation-flocculation is a typical process found in drinking water treatment plants (DWTPs) [54,55], it is also commonly employed in WWTPs. For example, Hidayaturrahman and Lee [49] reported removal efficiencies of MPs between 50–82% by means of a coagulation-flocculation process.

The effect of RSF in the MP elimination has been analysed in previous works with a wide variety of results. For example, in a WWTP located in Finland, MPs were reduced from 0.7 to 0.02 MPs/L, which means an efficiency of 97% [30]. In another study carried out in two German WWTPs, the use of a sand filter achieved a noteworthy MP removal (above 99%) [56], whereas Magni et al. [19] described a MP elimination by a RSF of only 50%.

The overall MP removal efficiency of the WWTP analysed in this work was between 89% and 95%, with an average value of 92.9 ± 2.1% and it is remarkable that no noticeable variation between months was detected, so rainfall and temperature does not seem to affect MP elimination. The removal efficiencies found in the facility analysed in the present work were within the range reported in different European WWTPs (72–98%) [3,57–60]. A wide variation can be found depending on the treatment technology used and the operating conditions in the WWTP [61], the origin and type of wastewater [20], as well as the sampling and identification methods used in the process, population density and regional development [40].

3.2. Characterization of MPs by Size, Shape and Colour

As explained before, the sampling procedure allowed the MP classification by size. According to Figure 3, on average, in influent samples MP ≥ 500 µm only accounted around 30% of total MPs, whereas 56% and 80% represented MPs higher than 250 µm and 100 µm, respectively. This indicates a major percentage of small MPs than usual in the influent since most WWTPs it has been reported MPs abundance with a size greater than 500 µm above 70% [18,45,46,61–63]. The variations in the percentages of MPs found in each range of size are noticeable thorough the treatment processes, i.e., the percentage of those MPs with a size greater than 500 µm decreased from 30% in the influent to 24% in the secondary effluent. At the same time, the percentage of the smallest particles (20–100 µm) increased from 20% in the influent to 23% in the secondary effluent. It should be noted that, after pre-treatment and secondary treatment, the MPs most easily eliminated were those larger than 500 µm (57%) and those with a size between 250–500 µm (52%), as can be seen in Table S2. This means that the grit and grease system and the secondary treatment removed the bigger MPs with higher efficiency than the smaller ones. Important variations in the percentages of the middle sizes were not detected and the sizes distribution in the final effluent is similar to the secondary one. In the final effluent samples, the vast majority of MPs were smaller than 500 µm, around 76%, whereas a quarter of the microplastics were smaller than 100 µm (Table S2). These results agree with other previous studies, which reported that most of the MPs in the final effluent were smaller than 500 µm. However, the percentage of MPs smaller than 100 µm in the effluents is usually over 60%, percentage higher than those found in this work [19,31,33,34,52,64–67]. Table S2 shows that, after tertiary treatment, the most easily eliminated MPs were, both, those larger than 500 µm and those with a size between 100–250 µm (approximately 30%). In addition, considering the temperature, it can be observed that in the warmest months (May–September) the MPs with a size higher of 250 µm presented abundances of 60–70%, while during those months with lower temperature (November, February–April), it is observed that the MPs with sizes less than 250 µm presented abundances of 60%. It has been reported that MP degradation are determined by the combined effect of different parameters, including temperature. Specifically, Ariza-Tarazona et al. [68] concluded that photolysis combined to low temperatures leads to plastic brittleness, which is in accordance with results commented above, since the coldest months showed a greater proportion of MPs smaller than 250 µm. Finally, it can be observed that the overall microplastic removal efficiency was higher in MPs larger than 500 µm (70%) compared to the rest of the sizes.

The morphological characteristics of MPs found in wastewater samples can be classified into five different types: fragments, fibres, microspheres or pellets, films and foams, as can be observed in Figure 4. Fragments exhibit irregular and opaque shapes, whereas fibres show a high length-width ratio. Pellets have spherical form, foams are fluffy particles and, finally, films have a relatively flat surface.

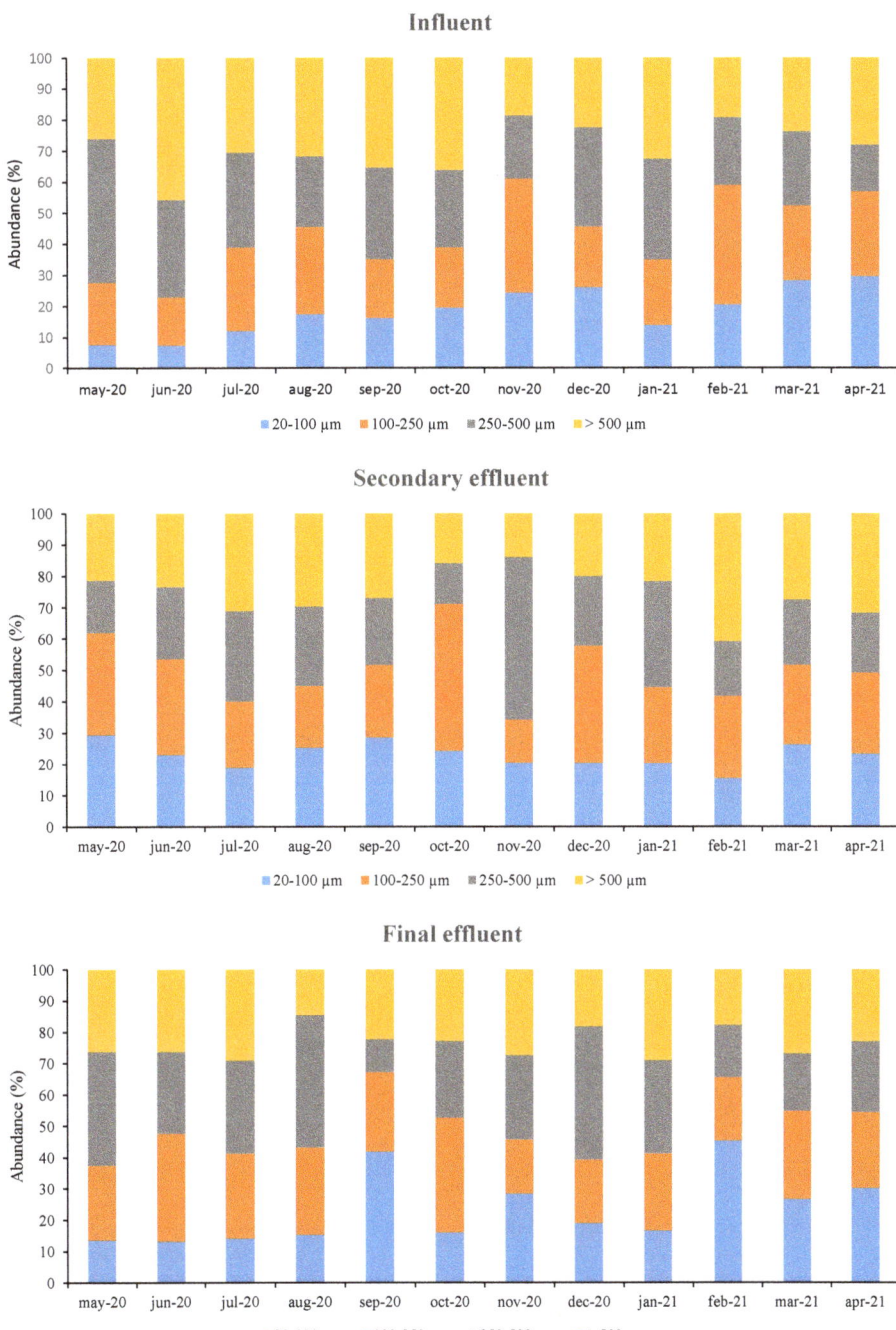

Figure 3. Size variation of microplastics in influent, secondary effluent, and final effluent samples during the period studied.

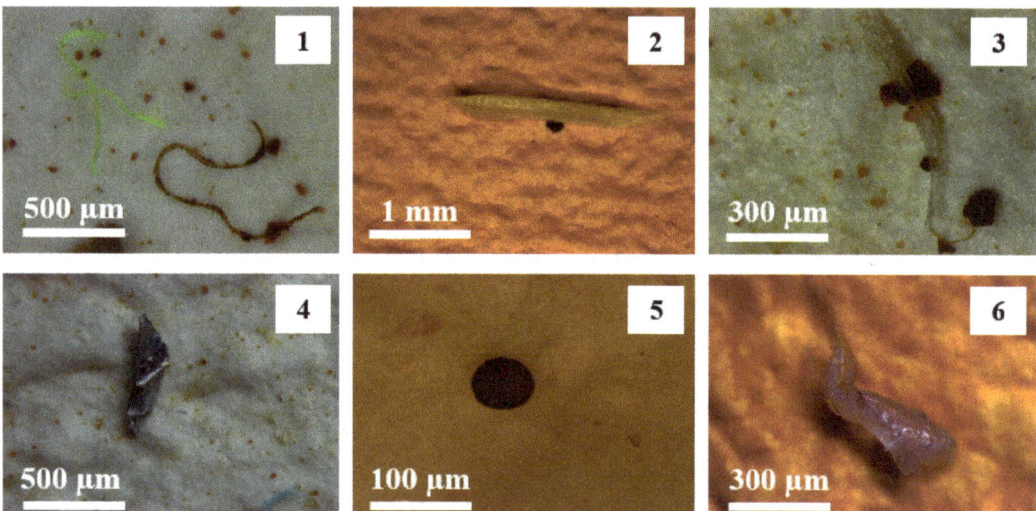

Figure 4. Examples of some microplastic particles found in this work and classified by shape and colour. (**1**): Green and yellow fibres, (**2**,**3**): Yellow and transparent fragments, (**4**): Grey film, (**5**): Black pellet, (**6**): Transparent foam.

Figure 5 shows the percentages of MP classified by shape found in the different points of sampling: influent, secondary effluent and final effluent. It can be observed that, in all samples, fibres and fragments constitute practically the totality of MPs (above 98%). According to the literature, fibres and fragments are the most predominant particle found in wastewater with a mean percentage of 56% and 34%, respectively [18,41,52,57,69–71]. Previous studies reported that fragments are the vast majority MPs [17,34,72,73], in concordance with the results obtained in this case study. Following the evolution of MPs through the wastewater treatment process (Figure 5), in influent samples the concentration of fragments and fibres ranged between 44.8–77.6% (with an average value of 64.9 ± 9.5%) and 20.0–55.2% (with an average value of 34.2 ± 10.2%), respectively. These percentages remained constant after the secondary treatment. However, in the final effluent samples, the concentration of fragments and fibres ranged between 46.1–81.4% and 18.6–61.0%, respectively, which shows a certain decrease in the abundance of fragments (with an average value of 57.3 ± 10.9%) and an increase in the percentage of fibres (with an average value of 40.3 ± 10.8%). This means that the tertiary treatment allowed a better removal of fragments (38%) than fibres (24%), as can be seen in Table S2. It has been reported that the high length-width ratio allows fibres to remain in water masses for more time than particles with other morphologies [2]. In addition, the overall removal efficiency shows a better elimination of fragments in comparison with fibres (67% vs. 56%). Finally, it is noteworthy that films, pellets and foams only account for 1–2%.

Respect to the MPs colour, white and black microparticles were the most common MPs at every sampling point, which means 81% of total MPs. The remaining percentage corresponds to red, blue, green, yellow and purple. This is agreement with previous studies that analyse MPs in WWTPs where higher abundances of white and black MPs were also detected [40,48,74].

Figure 5. Shape variation of microplastics in influent, secondary effluent and final effluent samples during the period studied.

3.3. Chemical Composition of MPs

The chemical composition is a relevant characteristic that determines the MP density and therefore, directly influences over the removal efficiency. Over 30 kinds of polymers have been described in wastewater samples of different WWTPs [51]. In this study PE, PP, PS, PA, PET and polyvinyl chloride (PVC) were detected in the wastewater samples (Figure 6). In the influent, on average, PP is the polymer most frequently detected with an abundance of 24.9 ± 5.5% (ranges between 15.8–37.4%), followed by PET with 23.2 ± 2.9% (27.8–18.1%), PE with 17.3 ± 4.2% (13.0–26.0%), PS with 14.5 ± 2.7% (10.4–17.3%), PA with 3.9 ± 3.4% (9.3–22.4%) and PVC with 6.2 ± 3.1% (1.5–10.7%). Different studies reported that most frequent polymers in urban wastewaters are PS (20–90%), PE (5–60%), PP (2–40%),

PET (3–38%), PA (2–35%) [2,3,75] and PVC in low abundances [61]. These data are in agreement with those found here, excepting for PS that was detected in percentages lower than values described in previous works. These variations in the abundance of different type of polymers in the influent are determined by the origin of wastewater that arrives to the WWTP (urban, industrial, agricultural) [74]. As the wastewater stream progresses through the different stages of WWTP, polymers less dense than wastewater, such as PP and PE, increased their proportion, being in the final effluent in percentages around 47.4 ± 3.6% and 29.6 ± 5.0%, respectively. On the contrary, polymers denser than wastewater, such as PS, PA, PET and PVC, decreased in abundance during the treatment processes due to their facility of settling, so they represented in the final effluent around 21%, whereas in the influent their proportion was notably higher (58%). An example of the FTIR spectra for each polymer are shown in Figure S2.

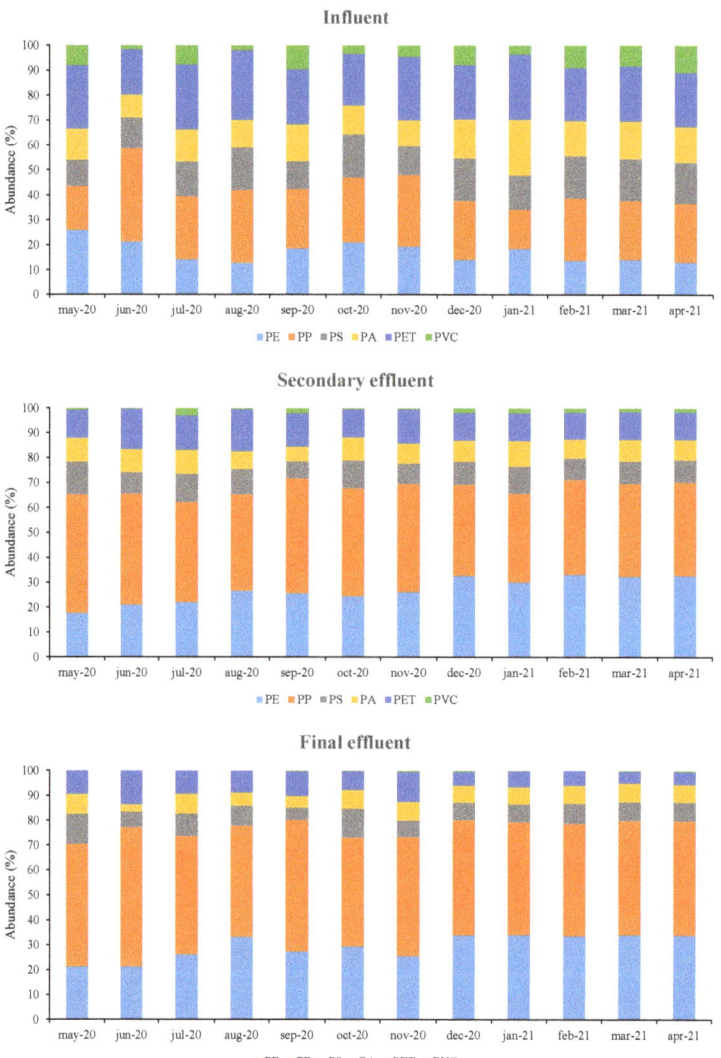

Figure 6. Composition variation of microplastics in influent, secondary effluent and final effluent samples during the period studied.

In addition, it has been analysed the relation between the chemical composition and the shape and colour of MPs during the different treatments in WWTP. Linking chemical composition with colour (Table S3), it is noticeable that, more than 90% of PVC microparticles were purple and yellow. Linking chemical composition with shape (Table S4), in the influent samples it has been found that PVC, PP, PET and PE have percentages of fragments of 98%, 77%, 74% and 67%, respectively. Moreover, 90% of the particles that corresponded to PA were fibres. Foams, pellets and films did not represent an abundance higher than 10% for any polymer.

3.4. Microplastics Entrapped in Sewage Sludge

It has been found that during the warmest months (May-September) the MP concentrations in dry sludge (28–39 MPs/g) were higher than those detected the rest of the months (12–22 MPs/g) (Figure 2c). These values are expressed by dry weight considering the sludge moisture of the different samples analysed (78–86% w/w). In the sludge samples there was a mean concentration of 24.0 ± 8.6 MPs/g dry sludge, value similar to those reported in literature for urban WWTPs [31,64,67,70,72]. According to Figure 2c, it can be observed that MP percentage retained in sludge varies between 47% and 100% with a mean value of 79%. These percentages are in agreement with those reported by different authors (8–92%) [16,28,32,76]. The removal of MPs in previous stages, i.e., during pre-treatment processes, can achieve notable values of elimination, for example, Murphy et al. [44] found that 45% of MPs that arrive at WWTP can be removed in grit and grease system. These percentages have been calculated based on the number of microplastics detected in the influent and the final effluent of the WWTP taking into account the daily flow (Table S5) in each sampling point. In addition, a trend between temperature and MPs retained in sewage sludge was observed, i.e., temperatures seem to favour the entrapping of MPs.

Physical and chemical properties of MPs retained in sludge samples were also analysed and results are summarised in Figure 7. Most MPs found in sludge are fragments and fibres ($57 \pm 18\%$ and $33 \pm 11\%$, respectively). Foams represent the 9%, but it should be noted that this specific shape was only detected in three samples (September, October and November) with percentages of 2%, 26% and 81%, respectively. The majority of the published works reported a higher abundance of fibres than fragments, with higher percentages than those found here (50–84%) [16,17,28,33,40,44,70,72,76]. However, it is remarkable that other works are in accordance with the results obtained in this case study, i.e., reported a higher proportion of fragments with respect to fibres [19,34,77].

As can be seen in Figure 7, no notable differences in abundance of MPs regarding chemical composition were found. The most predominant polymers in sludge samples were PET ($36 \pm 4\%$), followed by PS ($25 \pm 4\%$), PA ($20 \pm 4\%$) and PVC ($9 \pm 3\%$), in accordance with other studies, i.e., Kazour et al. [72] reported relative abundances of PS (25%), PET (20%), PA (10%) and PVC (5%) of the same order of magnitude than those found here. This agrees with the fact that the abundances of these polymers decreased throughout the wastewater treatment processes, as above commented. The high density of these polymers favours their sedimentation, being more easily entrapped in sludge. Regarding colour, around 82% of MPs found in sludge were white and black, as occurred in the wastewater samples.

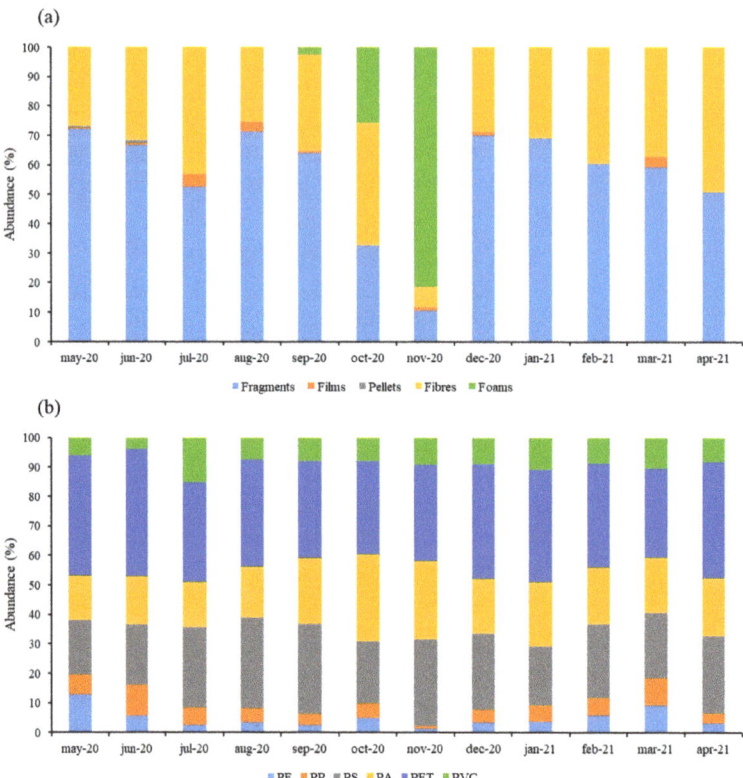

Figure 7. Abundance of microplastics in sludge samples according to (**a**) the shape and (**b**) chemical composition.

3.5. Release of MPs to the Environment

Several studies reported the environmental incidence of MPs emitted to the environment by WWTPs. As far as we know, until now, there have been seven works published that analysed the incidence of MPs in Spanish WWTPs [10,17,45,66,78–80]. The values found in the present work (average values of 16.1 ± 3.3 MPs/L and 1.1 ± 0.3 MPs/L in influent and effluent, respectively) were within the ranges reported by the previous works (between 2.7 MPs/L and 645 MPs/L in influent and 0.31 MPs/L and 16 MPs/L in effluent).

Considering the MP concentration detected in the influents (between 11.4 and 23.8 MPs/L) and the volume of wastewater that arrives at the WWTP (an average value between 4089 m^3/day and 5570 m^3/day) (Table S5), it can be estimated that between $5.57 \cdot 10^7$ and $1.27 \cdot 10^8$ microplastics enter into the facility each day. Since the removal efficiency of the studied facility is 92.9 ± 2.1%, approximately between $2.50 \cdot 10^6$ and $6.98 \cdot 10^6$ microplastics per day are emitted to the environment. For example, Edo et al. (2020) [17], who analysed a WWTP five times larger than that studied here, estimated that around $3 \cdot 10^8$ microplastics per day are release into the Henares River (Madrid), even though the WWTP reduce the MP concentration by 93%. This highlights the importance of WWTPs as source of MPs released into the environment.

In dry sludge samples, an average value of 24.0 ± 8.6 MPs/g is found, value lower than those values reported by other authors that analysed sewage sludge samples in the same country (Spain) (between 50 and 165 MPs/g) [17,22,78]. Considering that the MP concentration detected in the sludge (between 12.0 and 39.4 MPs per gram of dry sludge) and the kg of sludge generated in the WWTP (values between 1764 kg and 3976 kg)

(Table S6), it can be estimated that around between and $4.23 \cdot 10^7$ and $9.54 \cdot 10^7$ microplastics are entrapped in sludge. Thus, the subsequent management of the sludge is a determinant step to avoid the release of these MPs to the environment.

4. Conclusions

In this work the annual occurrence and fate of microplastics have been evaluated in a WWTP site in Southwest Europe employed as a case study. Results showed that this WWTP has a high removal efficiency (89–95%) all along the period studied, reducing considerably the number of MPs in treated water in comparison to influent values. Specifically, most microplastics (88%) were eliminated in the secondary treatment stage, being entrapped into the sludge. It was also found that the concentration of MPs in the influent was slightly higher during the warmer months (April–September) (17.1–23.8 MPs/L) compared to the colder ones (October–March) (11.4–15.6 MPs/L). MPs more easily eliminated from the wastewater samples were those with sizes greater than 500 µm and fragments and fibres were the shapes most frequently detected in wastewater and sludge samples. In addition, it was found that PP and PE were the commonest polymers in wastewater samples, whereas in sludge samples the majority were PET, PS and PA, which is due to the fact that denser polymers tend to settle more easily during the treatment processes. Furthermore, the temperature seems to favour the retention of MP in sludge. Future works should be focus on improving the removal of MPs from wastewater and, especially, from sewage sludge in order to reduce the release of these MPs to the environment.

Supplementary Materials: The following supporting information can be downloaded at: https://www.mdpi.com/article/10.3390/app12042133/s1, Figure S1: Device employed for wastewater sampling. The photograph shows the filtration module, pressure indicator and flow meter as part of the device, Figure S2: (a) Example of FTIR spectra registered for polyethylene (PE) obtained from the analysis of a black fibre recovered from the secondary effluent sample of July; (b): Example of FTIR spectra registered for polypropylene (PP) obtained from the analysis of a white fragment recovered from influent sample of August; (c): Example of FTIR spectra registered for polystyrene (PS) obtained from the analysis of a red fibre recovered from the influent sample of September; (d): Example of FTIR spectra registered for polyamide (PA) obtained from the analysis of a white foam recovered from the secondary effluent sample of October; (e): Example of FTIR spectra registered for polyethylene terephthalate (PET) obtained from the analysis of a white fragment recovered from the final effluent sample of March; (f): Example of FTIR spectra registered for polyvinyl chloride (PVC) obtained from the analysis of a black fragment recovered from the sludge sample of October, Table S1: Volumes of wastewater (L) and amount of sludge (g) sampled during the period of the study, Table S2: Size and shape evolution of microplastics after each treatment (influent, secondary effluent and final effluent) and the overall removal efficiency of each type of microplastic, Table S3: Relationship between the colours and chemical composition found for each sampling point expressed in percentage, Table S4: Relationship between the shapes and chemical composition found for each sampling point expressed in percentage, Table S5: Summary of the concentrations of microplastics (MPs/L) in the influent, secondary treatment and final effluent during the period of study. Influent and effluent average flow values are also indicated, Table S6: Summary of the concentrations of microplastics (MPs/g) in dehydrated sludge during the period of study. Average mass flow values are also indicated.

Author Contributions: Conceptualization, A.M.-M. and D.S.; methodology, A.M.-M., R.M.-D. and D.S.; validation, A.L. (Amanda Laca), A.L. (Adriana Laca) and M.D.; formal analysis, A.M.-M., D.S. and A.L. (Amanda Laca); investigation, A.M.-M., R.M.-D. and D.S.; resources, A.M.-M., D.S., A.M.-M. and M.D.; data curation, D.S., A.L. (Amanda Laca) and A.L. (Adriana Laca); writing—original draft preparation, A.M.-M. and D.S.; writing—review and editing, A.M.-M. and D.S.; visualization, A.L. (Amanda Laca) and A.L. (Adriana Laca); supervision, A.L. (Amanda Laca), A.L. (Adriana Laca) and M.D.; project administration, M.D. and A.R.; funding acquisition, A.R. All authors have read and agreed to the published version of the manuscript.

Funding: Foundation University of Oviedo, with Project FUO-395-19, has financed this work.

Institutional Review Board Statement: Not applicable.

Informed Consent Statement: Not applicable.

Data Availability Statement: Not applicable.

Acknowledgments: Authors gratefully acknowledge the cooperation of ACCIONA Agua and ESAMUR (Regional Entity for Sanitation and Wastewater Treatment of the Region of Murcia) with wastewater samples collection, especially to the workers of the WWTP of Caravaca de la Cruz.

Conflicts of Interest: The authors declare no conflict of interest.

Abbreviations

CAS	Conventional activated sludge
DM	Dynamic membranes
MBR	Membrane bioreactor
MP	Microplastic
PA	Polyamide
PE	Polyethylene
PET	Polyethylene terephthalate
PP	Polypropylene
PS	Polystyrene
PVC	Polyvinyl chloride
RSF	Rapid sand filtration
SBR	Sequencing batch reactor
WWTP	Wastewater treatment plant

References and Note

1. Masiá, P.; Sol, D.; Ardura, A.; Laca, A.; Borrell, Y.J.; Dopico, E.; Laca, A.; Machado-Schiaffino, G.; Díaz, M.; Garcia-Vazquez, E. Bioremediation as a promising strategy to microplastics removal in wastewater treatment plants. *Mar. Pollut. Bull.* **2020**, *156*, 111252. [CrossRef] [PubMed]
2. Ngo, P.L.; Pramanik, B.K.; Shah, K.; Roychand, R. Pathway, classification and removal efficiency of microplastics in wastewater treatment plants. *Environ. Pollut.* **2019**, *255*, 113326. [CrossRef]
3. Sol, D.; Laca, A.; Laca, A.; Díaz, M. Approaching the environmental problem of microplastics: Importance of WWTP treatments. *Sci. Total Environ.* **2020**, *740*, 140016. [CrossRef] [PubMed]
4. Boucher, J.; Friot, D. *Primary Microplastic in the Oceans: A Global Evaluation of Sources*; IUCN: Gland, Switzerland, 2017; 43p. [CrossRef]
5. De Falco, F.; Di Pace, E.; Cocca, M.; Avella, M. The contribution of washing processes of synthetic clothes to microplastic pollution. *Sci. Rep.* **2019**, *9*, 6633. [CrossRef] [PubMed]
6. De Falco, F.; Cocca, M.; Avella, M.; Thompson, R.C. Microfiber Release to Water, via Laundering, and to Air, via Everyday Use: A Comparison between Polyester Clothing with Differing Textile Parameters. *Environ. Sci. Technol.* **2020**, *54*, 3288–3296. [CrossRef]
7. Liu, X.; Yuan, W.; Di, M.; Li, Z.; Wang, J. Transfer and fate of microplastics during the conventional activated sludge process in one wastewater treatment plant of China. *Chem. Eng. J.* **2019**, *362*, 176–182. [CrossRef]
8. Rochman, C.M. Microplastics research—From sink to source. *Science* **2018**, *360*, 28–29. [CrossRef]
9. Ziajahromi, S.; Neale, P.A.; Leusch, F.D.L. Wastewater treatment plant effluent as a source of microplastics: Review of the fate, chemical interactions and potential risks to aquatic organisms. *Water Sci. Technol.* **2016**, *74*, 2253–2269. [CrossRef]
10. Bayo, J.; Olmos, S.; López-Castellanos, J. Microplastics in an urban wastewater treatment plant: The influence of physicochemical parameters and environmental factors. *Chemosphere* **2020**, *238*, 124593. [CrossRef]
11. Chen, G.; Feng, Q.; Wang, J. Mini-review of microplastics in the atmosphere and their risks to humans. *Sci. Total Environ.* **2020**, *703*, 135504. [CrossRef]
12. Prata, J.C. Airborne microplastics: Consequences to human health? *Environ. Pollut.* **2018**, *234*, 115–126. [CrossRef] [PubMed]
13. Liu, W.; Zhang, J.; Liu, H.; Guo, X.; Zhang, X.; Yao, X.; Cao, Z.; Zhang, T. A review of the removal of microplastics in global wastewater treatment plants: Characteristics and mechanisms. *Environ. Int.* **2021**, *146*, 106277. [CrossRef] [PubMed]
14. Habib, R.Z.; Thiemann, T.; Al Kendi, R. Microplastics and Wastewater Treatment Plants—A Review. *J. Water Resour. Prot.* **2020**, *12*, 1–35. [CrossRef]
15. Sol, D.; Laca, A.; Laca, A.; Díaz, M. Microplastics in Wastewater and Drinking Water Treatment Plants: Occurrence and Removal of Microfibres. *Appl. Sci.* **2021**, *11*, 10109. [CrossRef]
16. Gies, E.A.; LeNoble, J.L.; Noel, M.; Etemadifar, A.; Bishay, F.; Hall, E.R.; Ross, P.S. Retention of microplastic in a major secondary wastewater treatment plant in Vancouver, Canada. *Mar. Pollut. Bull.* **2018**, *133*, 553–561. [CrossRef]
17. Edo, C.; González-Pleiter, M.; Leganés, F.; Fernández-Piñas, F.; Rosal, R. Fate of microplastics in wastewater treatment plants and their environmental dispersion with effluent and sludge. *Environ. Pollut.* **2020**, *259*, 113837. [CrossRef]

18. Lares, M.; Ncibi, M.C.; Sillanpää, M. Occurrence, identification and removal of microplastic particles and fibers in conventional activated sludge process and advanced MBR technology. *Water Res.* **2018**, *133*, 236–246. [CrossRef]
19. Magni, S.; Binelli, A.; Pittura, L.; Avio, C.G.; Della Torre, C.; Parenti, C.C.; Gorbi, S.; Regoli, F. The fate of microplastics in an Italian Wastewater Treatment Plant. *Sci. Total Environ.* **2019**, *652*, 602–610. [CrossRef]
20. Mahon, A.M.; O'Connell, B.; Healy, M.G.; O'Connor, I.; Officer, R.; Nash, R.; Morrison, L. Microplastics in Sewage Sludge: Effects of Treatment. *Environ. Sci. Technol.* **2017**, *51*, 810–818. [CrossRef]
21. Rolsky, C.; Kelkar, V.; Driver, E.; Halden, R.U. Municipal sewage sludge as a source of microplastics in the environment. *Curr. Opin. Environ. Sci. Health* **2020**, *14*, 16–22. [CrossRef]
22. van den Berg, P.; Huerta-Lwanga, E.; Corradini, F.; Geissen, V. Sewage sludge application as a vehicle for microplastics in eastern Spanish agricultural soils. *Environ. Pollut.* **2020**, *261*, 114198. [CrossRef] [PubMed]
23. Auta, H.S.; Emenike, C.U.; Fauziah, S.H. Distribution and importance of microplastics in the marine environment: A review of the sources, fate, effects, and potential solutions. *Environ. Int.* **2017**, *102*, 165–176. [CrossRef]
24. Lu, L.; Luo, T.; Zhao, Y.; Cai, C.; Fu, Z.; Jin, Y. Interaction between microplastics and microorganism as well as gut microbiota: A consideration on environmental animal and human health. *Sci. Total Environ.* **2019**, *667*, 94–100. [CrossRef]
25. Wang, W.; Ge, J.; Yu, X. Bioavailability and toxicity of microplastics to fish species: A review. *Ecotoxicol. Environ. Saf.* **2019**, *189*, 109913. [CrossRef] [PubMed]
26. Xu, S.; Ma, J.; Ji, R.; Pan, K.; Miao, A.J. Microplastics in aquatic environments: Occurrence, accumulation, and biological effects. *Sci. Total Environ.* **2020**, *703*, 134699. [CrossRef] [PubMed]
27. Li, L.; Xu, G.; Yu, H.; Xing, J. Dynamic membrane for micro-particle removal in wastewater treatment: Performance and influencing factors. *Sci. Total Environ.* **2018**, *627*, 332–340. [CrossRef]
28. Lv, X.; Dong, Q.; Zuo, Z.; Liu, Y.; Huang, X.; Wu, W.M. Microplastics in a municipal wastewater treatment plant: Fate, dynamic distribution, removal efficiencies, and control strategies. *J. Clean. Prod.* **2019**, *225*, 579–586. [CrossRef]
29. Michielssen, M.R.; Michielssen, E.R.; Ni, J.; Duhaime, M.B. Fate of microplastics and other small anthropogenic litter (SAL) in wastewater treatment plants depends on unit processes employed. *Environ. Sci. Water Res. Technol.* **2016**, *2*, 1064–1073. [CrossRef]
30. Talvitie, J.; Mikola, A.; Koistinen, A.; Setälä, O. Solutions to microplastic pollution–removal of microplastics from wastewater effluent with advanced wastewater treatment technologies. *Water Res.* **2017**, *123*, 401–407. [CrossRef]
31. Lee, H.; Kim, Y. Treatment characteristics of microplastics at biological sewage treatment facilities in Korea. *Mar. Pollut. Bull.* **2018**, *137*, 1–8. [CrossRef]
32. Jiang, J.; Wang, X.; Ren, H.; Cao, G.; Xie, G.; Xing, D.; Liu, B. Investigation and fate of microplastics in wastewater and sludge filter cake from a wastewater treatment plant in China. *Sci. Total Environ.* **2020**, *746*, 141378. [CrossRef] [PubMed]
33. Naji, A.; Azadkhah, S.; Farahani, H.; Uddin, S.; Khan, F.R. Microplastics in wastewater outlets of Bandar Abbas city (Iran): A potential point source of microplastics into the Persian Gulf. *Chemosphere* **2021**, *262*, 128039. [CrossRef] [PubMed]
34. Pittura, L.; Foglia, A.; Akyol, Ç.; Cipolletta, G.; Benedetti, M.; Regoli, F.; Fatone, F. Microplastics in real wastewater treatment schemes: Comparative assessment and relevant inhibition effects on anaerobic processes. *Chemosphere* **2021**, *262*, 128415. [CrossRef]
35. Elkhatib, D.; Oyanedel-Craver, V. A critical review of extraction and identification methods of microplastics in wastewater and drinking water. *Environ. Sci. Technol.* **2020**, *54*, 7037–7049. [CrossRef]
36. Jung, M.R.; Horgen, F.D.; Orski, S.V.; Rodriguez, C.V.; Beers, K.L.; Balazs, G.H.; Jones, T.; Work, T.M.; Brignac, K.C.; Royer, S.J.; et al. Validation of ATR FT-IR to identify polymers of plastic marine debris, including those ingested by marine organisms. *Mar. Pollut. Bull.* **2018**, *127*, 704–716. [CrossRef]
37. Lin, L.; Zuo, L.Z.; Peng, J.P.; Cai, L.Q.; Fok, L.; Yan, Y.; Li, H.X.; Xu, X.R. Occurrence and distribution of microplastics in an urban river: A case study in the Pearl River along Guangzhou City, China. *Sci. Total Environ.* **2018**, *644*, 375–381. [CrossRef] [PubMed]
38. State Meteorological Agency. Territorial Meteorological Centre of Murcia.
39. Long, Z.; Pan, Z.; Wang, W.; Ren, J.; Yu, X.; Lin, L.; Lin, H.; Chen, H.; Jin, X. Microplastic abundance, characteristics, and removal in wastewater treatment plants in a coastal city of China. *Water Res.* **2019**, *155*, 255–265. [CrossRef]
40. Ren, P.; Dou, M.; Wang, C.; Li, G.; Jia, R. Abundance and removal characteristics of microplastics at a wastewater treatment plant in Zhengzhou. *Environ. Sci. Pollut. Res.* **2020**, *27*, 36295–36305. [CrossRef]
41. Yang, L.; Li, K.; Cui, S.; Kang, Y.; An, L.; Lei, K. Removal of microplastics in municipal sewage from China's largest water reclamation plant. *Water Res.* **2019**, *155*, 175–181. [CrossRef]
42. Hongprasith, N.; Kittimethawong, C.; Lertluksanaporn, R.; Eamchotchawalit, T.; Kittipongvises, S.; Lohwacharin, J. IR microspectroscopic identification of microplastics in municipal wastewater treatment plants. *Environ. Sci. Pollut. Res.* **2020**, *27*, 18557–18564. [CrossRef]
43. Magnusson, K.; Norén, F. *Screening of Microplastic Particles in and Down-Stream a Wastewater Treatment Plant*; C 55; IVL Swedish Environmental Research Institute: Stockholm, Sweden, 2014.
44. Murphy, F.; Ewins, C.; Carbonnier, F.; Quinn, B. Wastewater treatment works (WwTW) as a source of microplastics in the aquatic environment. *Environ. Sci. Technol.* **2016**, *50*, 5800–5808. [CrossRef] [PubMed]
45. Franco, A.A.; Arellano, J.M.; Albendín, G.; Rodríguez-Barroso, R.; Zahedi, S.; Quiroga, J.M.; Coello, M.D. Mapping microplastics in Cadiz (Spain): Occurrence of microplastics in municipal and industrial wastewaters. *J. Water Process. Eng.* **2020**, *38*, 101596. [CrossRef]

46. Cao, Y.; Wang, Q.; Ruan, Y.; Wu, R.; Chen, L.; Zhang, K.; Lam, P.K.S. Intra-day microplastic variations in wastewater: A case study of a sewage treatment plant in Hong Kong. *Mar. Pollut. Bull.* **2020**, *160*, 111535. [CrossRef]
47. Alavian Petroody, S.S.; Hashemi, S.H.; van Gestel, C.A.M. Factors affecting microplastic retention and emission by a wastewater treatment plant on the southern coast of Caspian Sea. *Chemosphere* **2020**, *261*, 128179. [CrossRef]
48. Ben-David, E.A.; Habibi, M.; Haddad, E.; Hasanin, M.; Angel, D.L.; Broth, A.M.; Sabbah, I. Microplastic distributions in a domestic wastewater treatment plant: Removal efficiency, seasonal variation and influence of sampling technique. *Sci. Total Environ.* **2021**, *752*, 141880. [CrossRef]
49. Hidayaturrahman, H.; Lee, T.G. A study on characteristics of microplastic in wastewater of South Korea: Identification, quantification, and fate of microplastics during treatment process. *Mar. Pollut. Bull.* **2019**, *146*, 696–702. [CrossRef] [PubMed]
50. Ruan, Y.; Zhang, K.; Wu, C.; Wu, R.; Lam, P.K. A preliminary screening of HBCD enantiomers transported by microplastics in wastewater treatment plants. *Sci. Total Environ.* **2019**, *674*, 171–178. [CrossRef]
51. Sun, J.; Dai, X.; Wang, Q.; van Loosdrecht, M.C.; Ni, B.J. Microplastics in wastewater treatment plants: Detection, occurrence and removal. *Water Res.* **2019**, *152*, 21–37. [CrossRef] [PubMed]
52. Ziajahromi, S.; Neale, P.A.; Rintoul, L.; Leusch, F.D. Wastewater treatment plants as a pathway for microplastics: Development of a new approach to sample wastewater-based microplastics. *Water Res.* **2017**, *112*, 93–99. [CrossRef]
53. Jiang, F.; Wang, M.; Ding, J.; Cao, W.; Sun, C. Occurrence and Seasonal Variation of Microplastics in the Effluent from Wastewater Treatment Plants in Qingdao, China. *J. Mar. Sci. Eng.* **2022**, *10*, 58. [CrossRef]
54. Shen, M.; Song, B.; Zhu, Y.; Zeng, G.; Zhang, Y.; Yang, Y.; Yi, H. Removal of microplastics via drinking water treatment: Current knowledge and future directions. *Chemosphere* **2020**, *251*, 126612. [CrossRef]
55. Novotna, K.; Cermakova, L.; Pivokonska, L.; Cajthaml, T.; Pivokonsky, M. Microplastics in drinking water treatment—Current knowledge and research needs. *Sci. Total Environ.* **2019**, *667*, 730–740. [CrossRef] [PubMed]
56. Wolff, S.; Weber, F.; Kerpen, J.; Winklhofer, M.; Engelhart, M.; Barkmann, L. Elimination of Microplastics by Downstream Sand Filters in Wastewater Treatment. *Water* **2021**, *13*, 33. [CrossRef]
57. Mason, S.A.; Garneau, D.; Sutton, R.; Chu, Y.; Ehmann, K.; Barnes, J.; Rogers, D.L. Microplastic pollution is widely detected in US municipal wastewater treatment plant effluent. *Environ. Pollut.* **2016**, *218*, 1045–1054. [CrossRef]
58. Takdastan, A.; Niari, M.H.; Babei, A.; Dobaradaran, S.; Jorfi, S.; Ahmadi, M. Occurrence and distribution of microplastic particles and the concentration of Di 2-ethyl hexyl phthalate (DEHP) in microplastics and wastewater in the wastewater treatment plant. *J. Environ. Manag.* **2021**, *280*, 111851. [CrossRef] [PubMed]
59. Yang, Z.; Li, S.; Ma, S.; Liu, P.; Peng, D.; Ouyang, Z.; Guo, X. Characteristics and removal efficiency of microplastics in sewage treatment plant of Xi'an City, northwest China. *Sci. Total Environ.* **2021**, *771*, 145377. [CrossRef]
60. Yuan, F.; Zhao, H.; Sun, H.; Zhao, J.; Sun, Y. Abundance, morphology, and removal efficiency of microplastics in two wastewater treatment plants in Nanjing, China. *Environ. Sci. Pollut. Res.* **2021**, *28*, 9327–9337. [CrossRef]
61. Wang, F.; Wang, B.; Duan, L.; Zhang, Y.; Zhou, Y.; Sui, Q.; Yu, G. Occurrence and distribution of microplastics in domestic, industrial, agricultural and aquacultural wastewater sources: A case study in Changzhou, China. *Water Res.* **2020**, *182*, 115956. [CrossRef]
62. Dris, R.; Gasperi, J.; Rocher, V.; Saad, M.; Renault, N.; Tassin, B. Microplastic contamination in an urban area: A case study in Greater Paris. *Environ. Chem.* **2015**, *12*, 592–599. [CrossRef]
63. Gündoğdu, S.; Çevik, C.; Güzel, E.; Kilercioğlu, S. Microplastics in municipal wastewater treatment plants in Turkey: A comparison of the influent and secondary effluent concentrations. *Environ. Monit. Assess.* **2018**, *190*, 626. [CrossRef]
64. Mintenig, S.M.; Int-Veen, I.; Löder, M.G.; Primpke, S.; Gerdts, G. Identification of microplastic in effluents of waste water treatment plants using focal plane array-based micro-Fourier-transform infrared imaging. *Water Res.* **2017**, *108*, 365–372. [CrossRef] [PubMed]
65. Simon, M.; van Alst, N.; Vollertsen, J. Quantification of microplastic mass and removal rates at wastewater treatment plants applying Focal Plane Array (FPA)-based Fourier Transform Infrared (FT-IR) imaging. *Water Res.* **2018**, *142*, 1–9. [CrossRef] [PubMed]
66. Franco, A.A.; Arellano, J.M.; Albendín, G.; Rodríguez-Barroso, R.; Coello, M.D. Microplastic pollution in wastewater treatment plants in the city of Cadiz: Abundance, removal efficiency and presence in receiving water body. *Sci. Total Environ.* **2021**, *776*, 145795. [CrossRef]
67. Xu, X.; Zhang, L.; Jian, Y.; Xue, Y.; Gao, Y.; Peng, M.; Jiang, S.; Zhang, Q. Influence of wastewater treatment process on pollution characteristics and fate of microplastics. *Mar. Pollut. Bull.* **2021**, *169*, 112448. [CrossRef]
68. Ariza-Tarazona, M.C.; Villarreal-Chiu, J.F.; Hernández-López, J.M.; De la Rosa, J.R.; Barbieri, V.; Siligardi, C.; Cedillo-González, E.I. Microplastic pollution reduction by a carbon and nitrogen-doped TiO_2: Effect of pH and temperature in the photocatalytic degradation process. *J. Hazard. Mater.* **2020**, *395*, 122632. [CrossRef]
69. Talvitie, J.; Heinonen, M.; Pääkkönen, J.P.; Vahtera, E.; Mikola, A.; Setälä, O.; Vahala, R. Do wastewater treatment plants act as a potential point source of microplastics? Preliminary study in the coastal Gulf of Finland, Baltic Sea. *Water Sci. Technol.* **2015**, *72*, 1495–1504. [CrossRef]
70. Tang, N.; Liu, X.; Xing, W. Microplastics in wastewater treatment plants of Wuhan, Central China: Abundance, removal, and potential source in household wastewater. *Sci. Total Environ.* **2020**, *745*, 141026. [CrossRef]
71. Xu, X.; Jian, Y.; Xue, Y.; Hou, Q.; Wang, L. Microplastics in the wastewater treatment plants (WWTPs): Occurrence and removal. *Chemosphere* **2019**, *235*, 1089–1096. [CrossRef]

72. Kazour, M.; Terki, S.; Rabhi, K.; Jemaa, S.; Khalaf, G.; Amara, R. Sources of microplastics pollution in the marine environment: Importance of wastewater treatment plant and coastal landfill. *Mar. Pollut. Bull.* **2019**, *146*, 608–618. [CrossRef]
73. Park, H.J.; Oh, M.J.; Kim, P.G.; Kim, G.; Jeong, D.H.; Ju, B.K.; Lee, W.S.; Chung, H.M.; Kang, H.J.; Kwon, J.H. National Reconnaissance Survey of Microplastics in Municipal Wastewater Treatments Plants in Korea. *Environ. Sci. Technol.* **2020**, *54*, 1503–1512. [CrossRef]
74. Conley, K.; Clum, A.; Deepe, J.; Lane, H.; Beckingham, B. Wastewater treatment plants as a source of microplastics to an urban estuary: Removal efficiencies and loading per capita over one year. *Water Res.* **2019**, *3*, 100030. [CrossRef] [PubMed]
75. Ali, I.; Ding, T.; Peng, C.; Naz, I.; Sun, H.; Li, J.; Liu, J. Micro- and nanoplastics in wastewater treatment plants: Occurrence, removal, fate impacts and remediation technologies—A critical review. *Chem. Eng. J.* **2021**, *423*, 130205. [CrossRef]
76. Ziajahromi, S.; Neale, P.A.; Silveira, I.T.; Chua, A.; Leusch, F.D.L. An audit of microplastic abundance throughout three Australian wastewater treatment plants. *Chemosphere* **2021**, *263*, 128294. [CrossRef] [PubMed]
77. Zhang, L.; Liu, J.; Xie, Y.; Zhong, S.; Gao, P. Occurrence and removal of microplastics from wastewater treatment plants in a typical tourist city in China. *J. Clean. Prod.* **2021**, *291*, 125968. [CrossRef]
78. Alvim, C.B.; Bes-Piá, M.A.; Mendoza-Roca, J.A. Separation and identification of microplastics from primary and secondary effluents and activated sludge from wastewater treatment plants. *Chem. Eng. J.* **2020**, *402*, 126293. [CrossRef]
79. Bayo, J.; López-Castellanos, J.; Olmos, S. Membrane bioreactor and rapid sand filtration for the removal of microplastics in an urban wastewater treatment plant. *Mar. Pollut. Bull.* **2020**, *156*, 111211. [CrossRef]
80. Bayo, J.; Olmos, S.; López-Castellanos, J. Assessment of Microplastics in a Municipal Wastewater Treatment Plant with Tertiary Treatment: Removal Efficiencies and Loading per Day into the Environment. *Water* **2021**, *13*, 1339. [CrossRef]

Review

Potential Use of Biochar in Pit Latrines as a Faecal Sludge Management Strategy to Reduce Water Resource Contaminations: A Review

Matthew Mamera [1,*], Johan J. van Tol [1], Makhosazana P. Aghoghovwia [1], Alfredo B. J. C. Nhantumbo [2], Lydia M. Chabala [3], Armindo Cambule [2], Hendrix Chalwe [3], Jeronimo C. Mufume [4] and Rogerio B. A. Rafael [2]

1. Department of Soil, Crop and Climate Sciences, Faculty of Natural Sciences, University of the Free State, Bloemfontein 9301, South Africa; vantoljj@ufs.ac.za (J.J.v.T.); AghoghovwiaMP@ufs.ac.za (M.P.A.)
2. Rural Engineering Department, Faculty of Agronomy and Forest Engineering, Eduardo Mondlane University, P.O. Box 257, Maputo 1102, Mozambique; abnhantumbo@yahoo.com (A.B.J.C.N.); armindo.cambule@uem.mz (A.C.); rogerborguete@gmail.com (R.B.A.R.)
3. Department of Soil Science, School of Agricultural Sciences, University of Zambia, P.O. Box 32379, Lusaka 10101, Zambia; lchabala@unza.zm (L.M.C.); hendrix.chalwe@unza.zm (H.C.)
4. Department of Morphological Sciences, Faculty of Medicine, Eduardo Mondlane University, P.O. Box 257, Maputo 1102, Mozambique; jmufume@gmail.com
* Correspondence: 2015319532@ufs4life.ac.za; Tel.: +27-67-090-4564

Abstract: Faecal sludge management (FSM) in most developing countries is still insufficient. Sanitation challenges within the sub-Saharan region have led to recurring epidemics of water- and sanitation-related diseases. The use of pit latrines has been recognised as an option for on-site sanitation purposes. However, there is also concern that pit latrine leachates may cause harm to human and ecological health. Integrated approaches for improved access to water and sanitation through proper faecal sludge management are needed to address these issues. Biochar a carbon-rich adsorbent produced from any organic biomass when integrated with soil can potentially reduce contamination. The incorporation of biochar in FSM studies has numerous benefits in the control of prospective contaminants (i.e., heavy metals and inorganic and organic pollutants). This review paper evaluated the potential use of biochar in FSM. It was shown from the reviewed articles that biochar is a viable option for faecal sludge management because of its ability to bind contaminants. Challenges and possible sustainable ways to incorporate biochar in pit latrine sludge management were also illustrated. Biochar use as a low-cost adsorbent in wastewater contaminant mitigation can improve the quality of water resources. Biochar-amended sludge can also be repurposed as a useful economical by-product.

Keywords: biochar; contaminants; pit latrines; sludge management; sustainable soil conditioner; water quality

1. Introduction

Faecal sludge management (FSM) in most developing countries of the sub-Saharan region is ineffective and insufficient, which causes a deepening of sanitation problems [1–4]. Improper pit emptying and sludge disposal have been attributed to factors such as shortages in suitable sanitation, poor drainage systems, and high groundwater fluctuations [1,5]. Further, sludge management is impacted by high transport and disposal costs in landfills. The permanent airspace disposals can also lead to human and environmental impacts [6]. Previous and recent latrine building projects have focused on constructing latrines without considering the emptying process and sludge management strategies [2].

Sanitation challenges within sub-Saharan Africa have led to recurring epidemics of sanitation-related diseases, including soil-transmitted helminth infections [7]. Outbreaks can occur periodically where water supplies and sanitation provisions are inadequate, most frequently in the developing world [3,4,8,9]. Between 1970 and 2011, African countries reported over 3 million suspected cholera cases, representing 46% of all cases reported globally [8]. Sub-Saharan Africa accounted for 86% of reported cases and 99% of deaths worldwide in 2011 [9–11]. Statistics of this nature are alarming and need urgent redress. While the reasons for these conditions are complex, part of the problem is the difficulty in accessing clean water and safe potable water, lack of sanitation, and the high costs involved. Pollution problems from pit latrines depend on climatic conditions, geological formations and soilscapes on the rate of soil contaminants migration. These factors lead to a need for scientific assessment of sludge management and pollution challenges. This can ensure that these sanitations are properly sited, designed, installed, monitored, and maintained [3,12]. Although the use of pit latrines as compared to open defecation can be beneficial, there are still concerns that they may cause dreadful human and ecological health impacts. This is associated with microbiological and chemical contamination of drinking water supplies through leaching into groundwater and surface water [11].

Integrated approaches for access and improvement of sanitation and water are needed to address these issues to curb the potential danger to public health and the environment. Creating simple and sustainable solutions for managing human excreta plays a direct role in slowing down the rate of environmental damage. This can be done by seeking alternative means that aim at reducing environmental pollution by faecal sludge, while not further depleting severely limited freshwater resources. Incorporation of soil a conditioner such as biochar has a high impact on the reduction in contaminant leaching [13–17].

Biochar is a high carbon-rich adsorbent produced from any organic biomass at high temperatures in conditions with limited oxygen [18]. Many studies to date have mostly focused on the potential of biochar to improve soil fertility for agricultural uses [19–21]. There are, however, numerous prospective benefits of integrating biochar in FSM studies. Such merits include: micro-organic mitigations [22]; reducing malodour [23]; contaminant barrier (bacteria and heavy metals) [15,24,25]; reduction in nitrogen [26], and carbon dioxide losses [27].

A gap in knowledge regarding the use of biochar to reduce the environmental threat of faecal sludge still exists. In the recent past, the potential of biochar to reduce leaching has been recognized, and several studies have been conducted on organic and inorganic pollution restriction by biochar. This review aimed to evaluate the potential of biochar in FSM through literature, which focused on biochar, sanitation, and faecal sludge studies. This review merits attention, because it explores alternative means for faecal sludge management, which can also be implemented in developing countries such as Mozambique, South Africa, and Zambia to minimize seepage of pit latrine waters and provide a sustainable soil conditioner for crop production.

2. On-Site Sanitation Systems

On-site sanitation is characterized by treatment and disposal of human waste, which is not removed to an off-site sanitation system [28,29]. Such sanitation facilities store wastes at the site of disposal, which decompose in situ [30]. These systems have two main categories; the wet latrines, which use water for flushing, and the dry latrines, which function without water sources. The different types of on-site sanitation systems [28,30,31] are pit latrines, ventilated improved pit latrines (VIPs), urine diversion (UD) toilets (Figure 1), ecological sanitation (EcoSan) latrines, Fossa Alterna, anaerobic biogas reactors, and septic tanks. A common pit latrine is composed of a simple top structure constructed over a pit and collects waste [32]. Improved pit latrines are a simple and low-cost type of sanitation system [13].

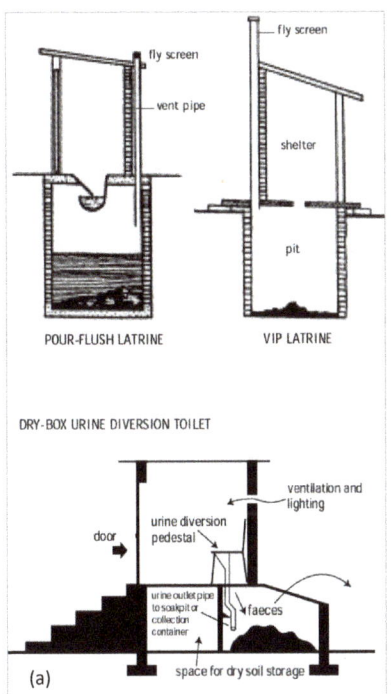

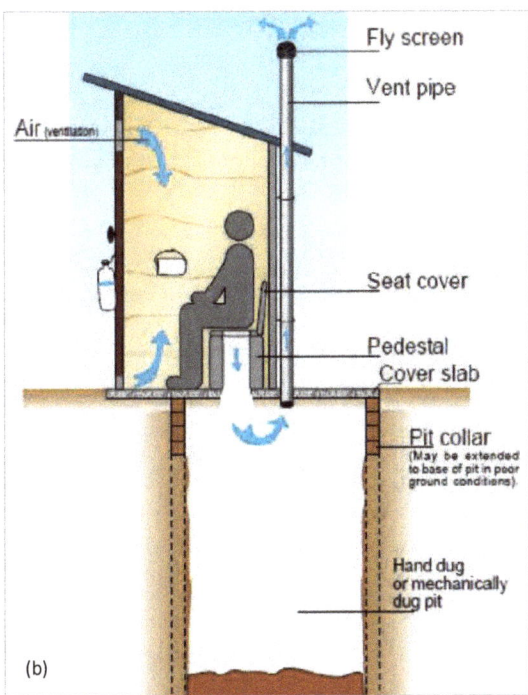

Figure 1. (**a**) Examples of on-site sanitation designs [30]; (**b**) typical structure for a VIP toilet system in South Africa [31] reproduced from the reference, copyright 2001, CC-BY-4.0.

2.1. Contamination Risks of Pit Latrines

On-site sanitation systems often represent a significant contamination threat towards groundwater associated with faecal matter accumulations, which can result in leaching of contaminants into the subsurface aquifer. Leachates in pits can lead to both microbiological and chemical contamination. In a pit latrine, the liquid fraction of waste that infiltrates into the soil is referred to as the hydraulic load [33]. Since pit latrines are usually not sealed [30], higher hydraulic loads can exceed natural attenuation potential in the sub-surface layers and cause direct contamination of groundwater sources. Designs of most pit latrines allow the liquid waste to infiltrate into the soil. Such wastes often contain micro-organisms and high nitrogen concentrations [30]. The hydrogeology in unlined pit latrines is extremely permeable, especially in coarser materials and fractured substratum. Such conditions promote rapid drainage in most natural soils [34]. Such designs of pit latrines allow for groundwater and surface water movements, which cause them to fill up rapidly [1,2]. Soil effluent infiltration rates of different soils not amended with carbon-based adsorbents such as biochar are shown in Table 1.

Table 1. Infiltration capacity of different soil types [33,35].

Type of Soil	Infiltration Capacity Settled Sewage (L per m^2 per Day)
Coarse or medium sand	50
Fine sand, loamy sand	33
Sandy loam, loam	25
Porous silty clay and porous clay loam	20
Compact silty loam, compact silty clay, loam and non-expansive clay	10
Expansive clay	<10

Pour-flush latrines have a much greater hydraulic load as compared to dry latrines; thus they have a higher contamination capacity [35]. Pit latrines normally are deeper than other on-site sanitations and tend to rely on infiltration of leachate through the surrounding soil [30]. Pit latrines pose a contamination risk to water sources such as wells nearby. Therefore, wells need to be well covered. Kiptum and Ndambuki [36] found a strong correlation between the types of well cover, with the one made of concrete being better than the one made of timber. Concrete covers guard the well against surface runoff and windblown substances and help to exclude spilled water.

2.2. On-Site Sanitation Waste Components and Health Risks

Human excreta are composed of several chemicals (Table 2) and pathogens (Table 3) species posing threats to human health and the natural environment. Nitrates and phosphates are a major concern. Higher concentrations of nitrates (>45 mg/L) in drinking water sources are harmful to humans [37–39]. One of the effects of human beings ingesting water with high concentrations of nitrates is methemoglobinemia or infantile cyanosis, i.e., "blue baby syndrome" in infants and oesophageal cancer in adults [40]. The probable long-term effects of nitrate pollutants should be included in the planning phase of sanitation programmes, as remedial action is challenging, and blending with low nitrate waters may be the only viable option [41]. High loading of phosphates in water sources results in eutrophication problems, having an impact on human well-being, social interaction, economic activities, and the natural environment [42].

The majority of studies that assessed microbiological quality of groundwater in relation to pit latrines used faecal indicator bacteria, i.e., total coliforms, faecal coliforms, and *E. coli* [8,43]. Bacterial pathogens cause some of the best known and most feared infectious diseases, such as cholera, typhoid, and dysentery, which still cause massive outbreaks of diarrhoeal disease and contribute to ongoing infections [44]. Their control in drinking water remains critical in all countries worldwide [43].

Table 2. Human waste composition [42,45,46].

Compound	Faeces (% Dry Weight)	Urine (% Weight)
Organic matter	88–97	65–85
Carbon (C)	44–55	11–17
Nitrogen (N)	5.0–7.0	15–19
Phosphorous (P_2O_5)	3.5–4.0	2.5–5.0
Potassium (K_2O)	1.0–2.5	3.0–4.5
Calcium (CaO)	4.5	4.5–6.0
Dry solids/person/day (g)	30–70	50–70

Table 3. Common bacteria and viruses found in human excreta as pathogenic contaminants [33].

Pathogen	Illness	Present in (Faeces/Urine)
Escherichia coli, Faecal coliforms	Diarrhoea	Both
Leptospira interrogans	Leptospirosis	Urine
Salmonella typhi	Typhoid	Both
Shigella spp.	Shigellosis	Faeces
Vibrio cholerae	Cholera	Faeces
Poliovirus	Poliomyelitis	Faeces
Rotaviruses	Enteritis	Faeces

2.3. Heavy Metal Composition of Faecal Sludge

The disposal of heavy metals remains as a major concern globally to water sources contamination [47]. Heavy metals in faecal effluent originate from natural and anthropogenic sources [48]. A substantial quantity of the anthropogenic releases of heavy metals accumulates in surface and groundwater ecosystems [49]. Industrial water treatment plant (IWTP) sludge has higher concentration of heavy metals as compared to other sources such as water treatment plants (WTP) and wastewater treatment plants (WWTP). Thus, they are mostly not recommended for soil amendment and ecological purposes [50,51]. Several industrial sectors contribute heavy metals to the environment through sludge disposals. Some of these sources include plants such as galvanic processes, dye productions, steel pickling, electroplating industry, and the recycling of lead batteries, among many others [50]. Heavy metals concentration in pit latrines is lower than reported in wastewater sludge [52]. However, heavy metal elements are one of the main persistent contaminants of pit latrine leaching or municipal wastewater [48,53]. The persistence of heavy metals in effluent is caused by their non-biodegradable and harmful nature [54]. Metals are mobilized and transported into the food web because of the leaching process from waste dumps, polluted soils, and water [55]. The most common toxic heavy metals in wastewater and sewage sludge include arsenic (As), lead (Pb), mercury (Hg), cadmium (Ca), chromium (Cr), copper (Cu), nickel (Ni), silver (Ag), and zinc (Zn) [48,53,56–59]. There is increasing evidence linking Hg, Pb, As, and Cd toxicants to the incidence of cognitive impairments and cancers in children [60]. Additionally, high concentrations of arsenic and other heavy metals can affect the nervous system and kidneys and may cause reproductive disorders, skin lesions, endocrinal damage, and vascular diseases [8,37].

2.4. Treatment of Faecal Sludge

In some developing countries still relying on pit latrines, filled up latrines are either closed or emptied and the sludge disposed off-site as waste [17]. Sludge can also be utilized as a soil ameliorant, for improving the soil status [61]. However, land application of sludge can also promote the pollution of water and soil by heavy metals [62]. Prior to sludge applications, conventional treatments are carried out [62], but that is not normally the case in most developing countries. The removal of heavy metal pollutants can be achieved through these conventional techniques to treat wastewater streams, including reduction or precipitation via chemical means, ion exchange, electro-chemical methods, and reverse osmosis. Nonetheless, such processes can be inadequate, especially for solutions with 1 to 100 (mg/L) of metal concentrations [63]. Other methods have also been successfully used for heavy metal removals, microbial remediation, and phytoremediation: cortex fruit wastes, including banana, kiwi, and tangerine peels [55]; activated carbon, peanut husk charcoal, fly ash, and natural zeolite [47]; composting and immobilization using biochar [64]. Such processes are cost effective, with non-hazardous end products [65]. The effective elimination of heavy metals from wastewater relies on several aspects, such as sludge concentration, the solubility of metal ions, pH, the metallic species and its concentration, and wastewater contamination load [63,66].

3. Biochar Adsorbents

Biochar is a material that has only recently been studied as an environmental amendment [16,67,68]. The use of biochar in pit latrine sludge treatment in most developing countries is still limited. This is primarily due to a lack of awareness in communities relying on pit latrines on contaminant immobilization potential of biochar [17,29]. Biochar applications historically predate several years in the Brazilian region, which led to development of "Terra Preta de Indio" soils [69]. Biochar has long been used to date archaeological deposits due to its persistence in the environment [70]. Within the past decade, biochar has been evaluated as a potential alternative to nutrient releases and leaching reduction from the soil [16]. However, the standard application rate of biochar for specific soils and crop combination to obtain the maximum positive results is not available yet [71].

Biochar is the by-product of any type of biomass that has undergone pyrolysis (see example in Figure 2) [20,72]. Pyrolysis is a process that changes biomass to a carbon-rich by-product as a result of the thermal degradation of organic materials by heating it to high temperatures in the absence of oxygen [70,73]. The pyrolysis process can be subdivided into separate categories: gasification (>800 °C), fast pyrolysis (~500 °C), and slow pyrolysis (450–650 °C) [74]. Slow pyrolysis is the best optimum pyrolysis process for the production of biochar [75,76]. The removal of volatile substances and the creation of crystalline carbons via condensations in biochar due to the increase in temperature from 400–500 °C enhance the adsorption abilities by generation of more pores [77,78]. Biochar is distinguishable from charcoal because of its usage as a soil amendment [21,79]. Responses of biochar are specific to the soil and climate within an area, biomass material, preparation method, and conditions [80,81]. Laird et al. [19] demonstrated that biochar carbon contents can range from <1% to >80% because of different biomass materials and pyrolysis conditions. Applied biochar in soils cannot be removed, so its use on a large scale has potential negative impacts on occupational health, environmental pollution, water quality, and food safety that need to be assessed [76].

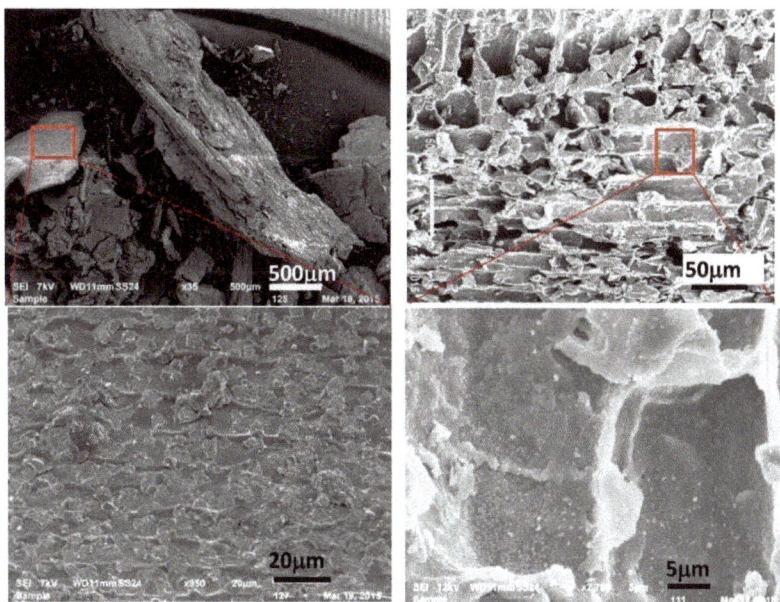

Figure 2. Scanning electron microscopy (SEM) pictures of pine sawdust biochar [72]. Reprinted from the reference with the permission, © 2021 John Wiley and Sons, Inc.

The char produced via pyrolysis is only known as biochar due to its amendment use in environmental management and production benefits to soil [18,73]. Bio-adsorbents similar to biochar are low cost with a high adsorption efficiency, as they require limited maintenance in wastewater contaminant treatments compared to other conventional methods [74–76]. The commercial worth of bio-adsorbents is low, and they are also accessible in abundance [76]. Affordability of an adsorbent can be increased, as they are stable and recyclable; hence, there is a high capacity for treatment of larger volumes of water contaminants over time [77]. Biochar's removal efficiencies for contaminants can be comparable to other commercial activated carbons because of improvements in cost-effective engineered biochar [78]. Biochar is also cheaper than other bio-adsorbents such as activated carbon, as it requires less production energy [74]. In addition to biochar's usage as a soil amendment, it is also used for carbon sequestration, mitigation of climate change, as a source of bio-energy, and waste management [18,70]. The high fraction of aromatic arrangements and high fraction of recalcitrant carbon (C) in biochar causes its resistance to chemical and biological degradation [82]. Biochar can persist in the soil for hundreds to thousands of years [70,83,84].

3.1. Properties of Biochar

Biochar characteristics such as the chemical composition, surface chemistry, particle and pore size distribution, and physical and chemical stabilization mechanisms in soils determine its effects on soil functions and faecal contaminants control [21]. Studies into biochar have demonstrated potentials for its use in increasing nutrient [19,70,85–88] and water retention [89–93] in soils, filtering heavy metals [94,95], reducing transport of microbes [14,15,22], increasing C sequestration [96–98], infiltration, soil aeration, root development, soil density, cation exchange capacity (CEC), and pH value [99–102]. The direct influence on soil structure, distribution of pore size, and density of the soil improves water holding capacity, aeration, and permeability [91,103,104].

Long-term properties including the stabilization of organic matter, slower release of nutrients from organic matter, and increased retention of cations have a huge impact to reducing the contamination of water resources [104,105]. Adsorption mechanisms studies showed that various types of interactions such as chemical bonding, chemical interaction, (complexation and/or precipitation), physical adsorption, ion exchange, and electrostatic attraction are largely responsible for binding faecal wastewater contaminants [22,94,95]. Physical sorption of metallic contaminants occurs on the surface area and pore volumes of biochar due to the high affinity of adsorption retained within the pores [74,106]. Most positively charged contaminants are sorbed through electrostatic attractions, ligands specificity, and several functional groups (e.g., hydroxyl, Alternariol-AOH, carboxylate, ACOOH) on biochar because of their negatively charged surfaces [107]. The effect can also promote surface complexities and precipitation of these contaminates to their physical mineral phases, which immobilizes them [108]. Physical or surface sorption also happens by diffusional movement of organic and inorganic elements into sorbent pores [74]. Contaminant sorption also occurs through the exchange of ionizable cations or protons and chemical bonding on the biochar surface with those species in solution. Furthermore, biochar's high pH influences adsorption, because it affects charges on the surface, levels of ionization, and speciation of the adsorbent [74,108]. These characteristics of biochar make it a viable soil and water quality amendment in studies associated with on-site sanitation systems and agricultural sludge usage. However, the effect of the ageing process on biochar properties has not been studied in detail; for example, adsorption capacities of biochar change with time [71].

3.2. Biochar Usage in Faecal Sludge Management

3.2.1. Nutrient Retention

Retention of soil nutrients has a direct effect to minimize risks of runoff and subsurface contamination of water bodies, highly reducing eutrophication and losses of nutrients [16]. Biochar can be a sustainable solution to latrine soil-bed nutrient leaching, consequently decreasing the nutrient concentrations in runoff and groundwater sources [16]. An increase in the CEC of a soil results in improved nutrient sorption on the colloids of biochar [19,20]. Biochar in soils also have the potential to largely decrease nitrogen losses and carbon dioxide releases [25]. Laird et al. [19] demonstrated an increase in N, organic C, P, K, Mg, and Ca in fine-loamy soil treated with hardwood biochar. A sorghum produced biochar also improved organic C and minimized greater losses of N, P, and K in overflow when combined with the soil [89]. Dissolved NO_3-N and PO_4-P decreased in wastewater bodies treated with a waste wood biochar-treated soil column [25] and also in soil mixed with an agricultural char (pecan, walnut, and coconut shells and rice hulls) [91]. Other than NO_3-N and PO_4-P, Beck et al. [92] observed a decrease in total N, total P, and total organic C. It has been seen that an increase in the application rates of biochar can also cause an increase in the nutrient holding capacity of the soil [20,90]. In their study, Huggins et al. [109] illustrated the efficiency of biochar to retain NH_4^+ and PO_4^{3-} from faecal wastewater (Table 4).

Table 4. Wastewater treatment and the retention ability of biochar [109].

Nutrient in Wastewater	Biochar Retention
NH_4^+ removal rate (g/m^3/d)	5.4 ± 0.51
NH_4^+ removal (%)	$90\% \pm 4\%$
PO_4^{3-} removal rate (g/m^3/d)	3.8 ± 0.01
PO_4^{3-} removal	$87\% \pm 2\%$

3.2.2. Heavy Metal Immobilization

Biochar can also act as a barrier to prevent heavy metals from percolating into groundwater aquifers and surface water resources [15,24,25]. Biochar has a high ability of filtering of heavy metals in contaminated soil and faecal sludge due to the potential of adsorbing metals on its surface [16]. Sequestration of Pb, Cd, Cu, and Ni has been reported from cottonseed hull biochar because of functional groups on the biochar surfaces [24]. As pH, volatile matter, O:C, and N:C ratios in the biochar increase, biochar's capacity to adsorb heavy metals also increases [16,24]. In a study using poultry litter biochar and green waste biochar, it was found that Cd and Pb elements in soil water decreased [94]. Conversely, Cu increased in the soil water because of more mobility through increased dissolved organic C [94]. Other studies observed no effect on Cu when the soil was mixed with a hardwood biochar produced at 750 °C [20]. Decreases in Cd, Cr, Cu, Pb, Ni, and Zn in excess faecal effluent leachates from a soil column amended with a wood-based biochar were also found [25]. Cu, Cd, and Pb were removed from aqueous solution after amending the soil with bamboo, sugarcane, hickory, and peanut hull biochars, with the bamboo biochar being most effective [62]. The high adsorption of heavy metals in the study by Zhou et al. [62] was associated with pH increases in the solution. An increase in Cu and Zn removal as pH increased has also been seen using hardwood and corn straw biochar produced at temperatures of 450 °C and 600 °C [110]. These findings are similar to Krueger et al. [111], showing the immobilization occurring when faecal sludge is treated with biochar (Figure 3). The long-term impacts of biochar on heavy metal elements sorption are limited and require more research due to the recalcitrance of biochar [16].

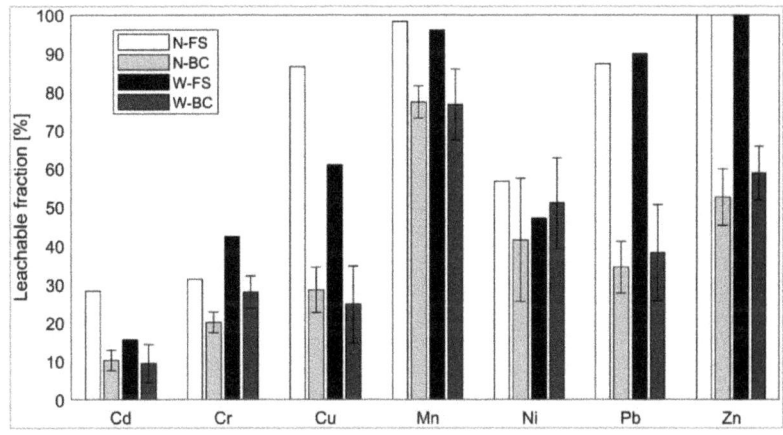

Figure 3. Mobility of heavy metals in faecal sludge (FS) and their derived biochars (BC); sludge sourced from Narsapur (N-FS) and Warangal (W-FS) Faecal Sludge Treatment Plants [111]. Reprinted from the reference with the permission, © 2019 Krueger et al., CC-BY-4.0.

3.2.3. Microbial Transport

The detection of *E. coli* and faecal coliforms (>1 CFU/100 mL) in soil and water resources above the guideline threshold [37–39] have a high risk on human health [16]. Presence of these bacteria indicates recent pollution from a faecal source such as pit latrine sanitations [13]. Bacteria can be infiltrated through the soil towards groundwater or move through overflow across the surface [16]. This threat to public health has urged investigation into microbial migration in soils, for which biochar amendments may be a solution [14,15].

When amended with soil, biochar can raise soil pH, which is essential for the mitigation of micro-organic pathogens such as *E. coli* and faecal coliform bacteria [22]. An increase in the organic matter, pH, conductivity, and dissolved organic C in a sandy soil using poultry manure biochar resulted in decreased soil *E. coli* and faecal coliforms migration [13]. Individually, these soil properties have been related to bacterial transport through soil [112–114]. Bolster and Abit [14] also demonstrated that biochar application rate, pyrolysis temperature, and *E. coli* surface properties largely contribute to the likely soil migrations. The higher temperature biochar (700 °C) exhibited a larger decrease in pathogen transport, possibly owing to the reduced negative surface charge of high-temperature biochars. The improved surface area of high-temperature biochars provides a higher adhesion of *E. coli* cells [15]. Biochar can also assist in dehydrating excreta because of its high water holding capacity, reducing malodour by adsorption [23] and thereby helping to keep insects such as flies away.

Biomass type used for biochar pyrolysis also plays a role in the transportation of soil *E. coli*, and faecal coliforms [16]. Comparison between poultry litter and pine chip biochars indicated that the internal pore structure of the woody biochar retained or adsorbed more bacteria [15]. Additionally, soils with higher clay contents have fewer detachments because of the electrostatic attraction force of the negatively charged microbes and the positive clay functional groups [15]. The influence of biochar on microbial movement through soil relies on biomass material, temperature, and soil texture [16]. Results from literature on the effect of biochar use in faecal sludge and contaminant reduction are presented in Table 5.

Table 5. Comparative literature values from previous faecal wastewater and sludge studies using biochar.

	Parameter	Type of Biochar	Concentration before Treatment	Concentration after Treatment	Literature
Nutrients	NH_3-N (mg/L)	Sludge and yellow pine biochar	2.8	3.0	Williams [16]
	NO_3-N (mg/L)	Sludge and yellow pine biochar	0.6	1.5	Williams [16]
	Nitrate (mg/L)	Effluent and waste wood pellets biochar	27	3.0	Reddy et al. [25]
	NH_4 (mg/L)	Wastewater and lodge pole pine wood biochar	50	5	Huggins et al. [109]
	Phosphate (mg/L)	Effluent and waste wood pellets biochar	0.57	0.4	Reddy et al. [25]
	Phosphate (mg/L)	Wastewater and lodge pole pine wood biochar	18	2	Huggins et al. [109]
	Phosphate (mg/L)	Faecal sludge and biochar	31	6.2	Krueger et al. [111]
Bacteria	E. coli (MPN/100 mL)	Effluent and waste wood pellets biochar	7400	5000	Reddy et al. [25]
	E. coli (MPN/100 mL)	Waste effluent and Monterey pine + eucalyptus biochar	291	<1	Kranner et al. [115]
	E. coli (CFU/100 mL)	Effluent and poultry litter and pine chips biochar	87	1.6	Abit et al. [15]
	Faecal coliforms (MPN/100mL)	Sludge and yellow pine biochar	150	26	Williams [16]
	Enterococci (MPN/100mL)	Waste effluent and Monterey pine + eucalyptus biochar	146	1	Kranner et al. [115]
	Arsenic (mg/L)	Wastewater and lodge pole pine wood biochar	27.9	0.01	Huggins et al. [109]
	Cadmium (mg/L)	Effluent and waste wood pellets biochar	24	17	Reddy et al. [25]
Heavy metals	Cadmium (ppm)	Aqueous concentrations + bamboo, sugarcane bagasse, hickory wood, and peanut hull biochars	30	<1	Zhou et al. [62]
	Cadmium (mg/L)	Wastewater and lodge pole pine wood biochar	11.1	<1	Huggins et al. [109]
	Cadmium (mg/L)	Faecal sludge and biochar	13.5	1.2	Krueger et al. [111]
	Chromium (mg/L)	Effluent and waste wood pellets biochar	5.13	5	Reddy et al. [25]
	Chromium (mg/L)	Wastewater and lodge pole pine wood biochar	34	0.1	Huggins et al. [109]
	Chromium (mg/L)	Faecal sludge and biochar	56.1	19	Krueger et al. [110]
	Copper (mg/L)	Effluent and waste wood pellets biochar	5	0.12	Reddy et al. [25]
	Copper (ppm)	Aqueous concentrations + bamboo, sugarcane bagasse, hickory wood, and peanut hull biochars	30	<1	Zhou et al. [62]
	Copper (mg/L)	Wastewater and lodge pole pine wood biochar	8.3	0.04	Huggins et al. [109]
	Copper (mg/L)	Faecal sludge and biochar	463	209	Krueger et al. [111]
	Lead (mg/L)	Effluent and waste wood pellets biochar	0.48	<0	Reddy et al. [25]
	Lead (ppm)	Aqueous concentrations + bamboo, sugarcane bagasse, hickory wood, and peanut hull biochars	50	<1	Zhou et al. [62]
	Lead (mg/L)	Wastewater and lodge pole pine wood biochar	13.5	<1	Huggins et al. [109]
	Nickel (mg/L)	Effluent and waste wood pellets biochar	110.61	80	Reddy et al. [25]
	Zinc (mg/L)	Effluent and waste wood pellets biochar	0.86	0.5	Reddy et al. [25]
	Zinc (mg/L)	Wastewater and lodge pole pine wood biochar	90.4	0.03	Huggins et al. [109]
Malodour	Malodour reconstitution solution (ORS) (O.U./m³)	Malodour solution + bamboo char, faecal char, and pine char	173	73	Stetina [23]
	ORS+H_2S (O.U./m³)	Malodour solution + bamboo char, faecal char, and pine char	181	49	Stetina [23]
	Butyric acid (O.U./m³)	Malodour solution + bamboo char, faecal char, and pine char	15	7	Starkenmann et al. [70]; Stetina [23]
	Indole (O.U./m³)	Malodour solution + bamboo char, faecal char, and pine char	23	12	Stetina [23]
	p-Cresol (O.U./m³)	Malodour solution + bamboo char, faecal char, and pine char	15	7	Starkenmann et al. [70]; Stetina [23]

4. Challenges, Sustainability, and Potential in Application of Biochar in Sludge Management

The most common challenge within communities using pit latrines is the ethical norm on the acceptance to repurpose biochar-treated faecal sludge for crop production [17]. Even though the biochar-treated sludge by-product can have acceptable threshold levels for most heavy metals and inorganic and organic contaminants, societies and communities in most developing countries treat human sludge as undesirable waste. In addition, the lack of knowledge in the biochar production process remains a challenge. The International Biochar Initiative [73] set guidelines on standards for production of biochar for use as soil amendments. However, information on biochar production for use in the treatment of contaminates and faecal sludge is limited. Moreover, most communities using latrines have livestock, which relies on the biomass material also needed to produce biochar. Nonetheless, the use of biochar in pit latrine sludge management can also be made sustainable, as the production process of biochar is regarded as an efficient management method to dispose of many organic wastes. However, advantages and disadvantages between the economic cost (production) and benefit value (application) of biochar need to be carefully measured. In addition, to enhance economic availability, easier production processes and cheaper sources of raw biomass materials need to be discovered to enhance economic availability [116]. Heavy metals can contaminate faecal sludge if toilets are also used to dispose of materials other than faecal sludge [52]. Studies have reported that for any new technology to be successfully integrated in a society, community awareness and engagement is important [13,17,24].

Education on the appropriate use of toilets is important [52], and application of biochar in latrines can be viable since a typical standard pit latrine only measures an approximate pit area of 2 m × 2 m [31,42,61]. In comparison to uses for amendment purposes in soil fertility and agriculture, sludge treatment can be more cost-effective, as the required biochar quantities are less bulky. The use of biochar has also been proven to increase faecal sludge decomposition, which can reduce the pit filling rates and increase the lifespan of a latrine. Biochar is also now commercially produced, which can also increase accessibility for sludge treatment and management uses. The high adsorption properties of biochar for water pollutants can assist with in situ sorbent and faecal sludge treatments. Such low-cost adsorbents can improve water quality through contaminant management.

5. Conclusions

This review focused on the potential uses of biochar in faecal sludge management (FSM) practices in most developing countries relying on pit latrine sanitation systems. Initially, the designs of pit latrines and the potential ways pollutants may migrate towards water resources without biochar amendments were outlined from previous literature. To understand the pollutant pit latrine leaching threat, the composition (heavy metals and inorganic and organic contaminates) of the stored in situ faecal excreta is important. Possible ways were explored on the effectiveness of biochar use in aqueous waste contaminant adsorption. The physical and chemical properties of biochar mostly determine its adsorption ability as an adsorbent in faecal waste management. Potential challenges in the adoption of biochar in faecal sludge management were also reviewed. Motivation can also be necessary to encourage communities using latrines to adopt biochar as an alternative low-cost carbon-rich absorbent for faecal sludge treatment. Biochar has high potential to effectively treat faecal sludge and control the migration of pit latrine pollutants.

Future Research Perspectives in Faecal Sludge Management

Studies on characterization and potential applications of biochar for several uses have been performed and research gaps have been indicated. Potential research lines that can be summarized as: (i) focus on potential secondary ecological risks in the process of biochar production by screening and pre-treating raw material to remove pollutants derived from biomass [100]; (ii) long-term stability and effect of biochar on agricultural soil characteristics; (iii) assessment of multifunctional biochar materials on multi-heavy metals contaminated soils and as engineering application; (iv) cost-benefit analysis to enhance economic availability to improve production efficiency and reduce economic constraints [63,100]; (v) soil toxicity to organisms and plants induced by biochar [63]. These research gaps are also applicable for biochar use in faecal sludge management. The integration of biochar on faecal sludge management has the potential to be adopted by smallholder farmers who have limited access to fertilizers due to financial limitations. The success for this integration requires additional detailed studies considering that the smallholder farmers have more possibilities to produce charcoal than biochar:

- Similarities on biochar and charcoal application on faecal sludge management;
- Assess the effectiveness of production and use of biochar (or charcoal) as low-cost faecal treatment techniques to contain pollutants (i.e., heavy metals and microbial elements) based on dominant plant species to resolve the large variations in environments and respective mechanism;
- Assess the potential use of faecal sludge treated with biochar as soil amendment for nutrients sources and the risk to release heavy metals in agricultural production;
- The long-term impacts of biochar (or charcoal) recalcitrance in faecal sludge on heavy metal elements retention and nutrient release in agricultural production;
- Socio-economic benefits from use of faecal sludge amended with biochar as soil fertilizer for agricultural production;
- Potential retention of pollutants by biochar and/or charcoal from hydraulic loads in dry pit latrines;

Author Contributions: Conceptualization, M.M., J.J.v.T. and M.P.A.; methodology, M.M., J.J.v.T., M.P.A., A.B.J.C.N., L.M.C. and A.C.; validation, J.J.v.T., M.P.A., A.B.J.C.N., L.M.C., A.C., H.C., J.C.M. and R.B.A.R.; investigation, M.M., J.J.v.T., M.P.A., A.B.J.C.N., L.M.C., A.C., H.C., J.C.M. and R.B.A.R.; resources, J.J.v.T., L.M.C. and A.C.; data curation, M.M.; writing—original draft preparation, M.M.; writing—review and editing, M.M., J.J.v.T., M.P.A., A.B.J.C.N., L.M.C., A.C., H.C., J.C.M. and R.B.A.R.; supervision, J.J.v.T. and M.P.A.; project administration, J.J.v.T., L.M.C. and A.C.; funding acquisition, J.J.v.T., L.M.C. and A.C. All authors have read and agreed to the published version of the manuscript.

Funding: This research was funded by the National Research Fund (NRF) and South Africa–Mozambique–Zambia NRF Trilateral Joint Research (ZAM180911357528-118479).

Institutional Review Board Statement: The study was conducted according to the guidelines of the Declaration of Helsinki, and approved by the General/Human Research Ethics Committee (GHREC)-N.o-UFSHSD2019/1012. Environmental and Biosafety Research Ethics Committee (EBREC)-N.o-UFS-ESD2019/0066, University of the Free State, SA.

Informed Consent Statement: Informed consent was obtained from all subjects involved in the study. Written informed consent has been obtained from the patient(s) to publish this paper.

Conflicts of Interest: The authors declare no conflict of interest.

References

1. Amaka, G.; Costa, C.; Baghirathan, R.V. Working small sanitation enterprise for faecal sludge management in peri-urban Maputo: The experience of WSUP. *Sanit. Matters* **2011**, *4*, 6.
2. WRC WIN-SA. Faecal sludge management in Africa developments, research and innovations. *Sanit. Matters* **2013**, *4*, 6–28.
3. Armah, F.A.; Ekumah, B.; Yawson, D.O.; Odoi, J.O.; Afitiri, A.R.; Nyieku, F.E. Access to improved water and sanitation in sub-Saharan Africa in a quarter century. *Heliyon* **2018**, *4*, e00931. [CrossRef] [PubMed]
4. WHO/UNICEF. The Joint Monitoring Programme (JMP) Report—Progress on Household Drinking Water, Sanitation and Hygiene 2000–2020. Five Years into the SDGs. 2021. Available online: https://www.unicef.org/press-releases/billions-people-will-lack-access-safe-water-sanitation-and-hygiene-2030-unless (accessed on 30 October 2021).
5. Capone, D.; Buxton, H.; Cumming, O.; Dreibelbis, R.; Knee, J.; Nalá, R.; Ross, I.; Brown, J. Impact of an intervention to improve pit latrine emptying practices in low income urban neighborhoods of Maputo, Mozambique. *Int. J. Hyg. Environ. Health* **2020**, *226*, 113480. [CrossRef]
6. Harrison, J.; Wilson, D. Towards sustainable pit latrine management through LaDePa. *Sustain. Sanit. Pract.* **2012**, *13*, 25–32.

7. Pullan, R.L.; Brooker, S.J. The global limites and population at risk of soil-transmitted helminth infections in 2010. *Parasites Vectors* **2012**, *5*, 81.
8. World Health Organization. Diarrheal Disease Fact Sheet. 2018. Available online: http://www.who.Int/mediacentre/factsheets/fs330/en/ (accessed on 12 March 2020).
9. Statistics South Africa. *Sustainable Development Goals (SDGs): Country Report*; Statistics South Africa: Pretoria, South Africa, 2019; 316p, ISBN 978-0-621-47619-4.
10. World Health Organization. Cholera Annual Report. *Wkly. Epidemiol. Rec.* **2020**, *95*, 441–448.
11. Mengel, M.A.; Delrieu, I.; Heyerdahl, L.; Gessner, B.D. Cholera outbreaks in Africa. *Curr. Top. Microbiol. Immunol.* **2014**, *379*, 44–117. [CrossRef]
12. Lorentz, S.A.; Wickham, B.; Ismail, S.; Rodda, N.; van Tol, J.J.; Cobus, C.C.; Kunene, B. Investigation into Pollution from On-Site Dry Sanitation Systems. Deliverable 6: Progress Report II on Field Evaluations. In *Water Research Commission Project K5/2115*; Water Research Commission: Pretoria, South Africa, 2013.
13. Graham, J.P.; Polizzotto, M.L. Pit Latrines and Their Impacts on Groundwater Quality: A Systematic Review. *Environ. Health Perspect.* **2013**, *121*, 521–530.
14. Bolster, C.H.; Abit, S.M. Biochar Pyrolyzed at Two Temperatures Affects *Escherichia coli* Transport through a Sandy Soil. *J. Environ. Qual.* **2012**, *41*, 124–133. [CrossRef]
15. Abit, S.M.; Bolster, C.H.; Cantrell, K.B.; Flores, J.Q.; Walker, S.L. Transport of *Escherichia coli*, *Salmonella typhimurium*, and Microspheres in Biochar-Amended Soils with Different Textures. *J. Environ. Qual.* **2014**, *43*, 371–378. [PubMed]
16. Williams, R. Effectiveness of Biochar Addition in Reducing Concentrations of Selected Nutrients and Bacteria in Runoff. Master's Thesis, University of Kentucky, Lexington, KY, USA. Available online: https://uknowledge.uky.edu/bae_etds/40 (accessed on 11 October 2021).
17. Mamera, M.; van Tol, J.J.; Aghoghovwia, M.P.; Mapetere, G.T. Community Faecal Management Strategies and Perceptions on Sludge Use in Agriculture. *Int. J. Environ. Res. Public Health* **2020**, *17*, 4128. [CrossRef]
18. Lehmann, J.; Joseph, S. Biochar for environmental management: An introduction. In *Biochar for Environmental Management: Science and Technology*; Lehmann, J., Joseph, S., Eds.; Earth Scan: London, UK, 2009; pp. 1–12.
19. Laird, D.A.; Fleming, P.; Davis, D.D.; Horton, R.; Wang, B.Q.; Karlen, D.L. Impact of biochar amendments on the quality of a typical midwestern agricultural soil. *Geoderma* **2010**, *158*, 443–449.
20. Mackie, K.A.; Marhan, S.; Ditterich, F.; Schmidt, H.P.; Kandeler, E. The effects of biochar and compost amendments on copper immobilization and soil microorganisms in a temperate vineyard. *Agric. Ecosyst. Environ.* **2015**, *201*, 58–69. [CrossRef]
21. Nyambo, P.; Taeni, T.; Chiduza, C.; Araya, T. Effects of Maize Residue Biochar Amendments on Soil Properties and Soil Loss on Acidic Hutton Soil. *Agronomy* **2018**, *8*, 256. [CrossRef]
22. Mohanty, S.K.; Cantrell, K.B.; Nelson, K.L.; Boehm, A.B. Efficacy of biochar to remove *Escherichia coli* from stormwater under steady and intermittent flow. *Water Resour.* **2014**, *61*, 288–296. [CrossRef] [PubMed]
23. Stetina, K. Control of Fecal Malodor by Adsorption onto Biochar. Ph.D. Thesis, University of Colorado, Boulder, CO, USA, 2017; pp. 58–69.
24. Uchimaya, M.; Wartelle, L.H.; Klasson, K.T.; Fortier, C.A.; Lima, I.M. Influence of Pyrolysis Temperature on Biochar Property and Function as a Heavy Metal Sorbent in Soil. *J. Agric. Food Chem.* **2011**, *59*, 2501–2510.
25. Reddy, K.R.; Xie, T.; Dastgheibi, S. Evaluation of Biochar as a Potential Filter Media for the Removal of Mixed Contaminants from Urban Storm Water Runoff. *J. Environ. Eng.* **2014**, *140*, 04014043. [CrossRef]
26. Aghoghovwia, M.P.; Hardie, A.G.; Rozanov, A.B. Characterisation, adsorption and desorption of ammonium and nitrate of biochar derived from different feedstocks. *Environ. Technol.* **2020**, *10*, 180. [CrossRef]
27. Aghoghovwia, M.P. Effect of Different Biochars on Inorganic Nitrogen Availability. Ph.D. Thesis, University of Stellenbosch, Stellenbosch, South Africa, 2018.
28. Tissington, K. *Basic Sanitation in South Africa: A Guide to Legislation, Policy and Practice*; Socio-Economic Rights Institute of South Africa (SERI): Johannesburg, South Africa, 2011.
29. Mamera, M.; van Tol, J.J. Application of Hydropedological Information to Conceptualize Pollution Migration from Dry Sanitation Systems in the Ntabelanga Catchment Area, South Africa. *Air Soil Water Res.* **2018**, *11*, 1178622118795485.
30. ARGOSS. *Guidelines for Assessing the Risk to Groundwater from On-Site Sanitation*; BGS Commissioned Report CR/01/142; British Geological Survey: Wallingford, UK, 2001.
31. DWAF. Sanitation Technology Options. *RSA Rep.* **2002**, *5*, 9–13.
32. Department of Environmental Affairs (DEA). National Policy for the Provision of Basic Refuse Removal Services to Indigent Households. 2010. Available online: https://www.environment.gov.za/sites/default/files/gazetted_notices/np_basicrefuse_removalservices_indigenthouseholds (accessed on 21 March 2021).
33. Franceys, R.; Pickford, J.; Reed, R. *A Guide to the Development of On-Site Sanitation*; World Health Organization: Geneva, Switzerland, 1992; Available online: http://www.who.int/water_sanitation_health/hygiene/envsan/onsitesan.pdf (accessed on 15 June 2019).
34. Chirwa, C.F.C.; Hall, R.P.; Krometis, L.H.; Vance, E.A.; Edwards, A.; Guan, T.; Holm, R.H. Pit latrine fecal sludge resistance using a dynamic cone penetrometer in low income areas in Mzuzu City, Malawi. *Int. J. Environ. Res. Public Health* **2017**, *14*, 87. [CrossRef]

35. Howard, B.; Reed, D.; McChesney, T.; Taylor, R. Human excreta and sanitation: Control and protection. In *Protecting Groundwater for Health: Managing the Quality of Drinking-Water Sources*; IWA Publishing: London, UK, 2006; ISBN 1843390795.
36. Kiptum, C.K.; Ndambuki, J.M. Well water contamination by pit latrines: A case study of Langas. *Int. J. Water Resour. Environ. Eng.* **2012**, *4*, 35–43. [CrossRef]
37. WHO. *Guidelines for Drinking-Water Quality*, 3rd ed.; World Health Organization: Geneva, Switzerland, 2008; Volume 1.
38. USEPA. *National Primary Water Drinking Regulations*; United States Environmental Protection Agency: Washington, DC, USA, 2009. Available online: http://www.epa.gov/ogwdw/consumer/pdf/mcl.pdf (accessed on 15 June 2019).
39. SANS. *South African National Standard: Drinking Water Part 1: Microbial, Physical, Aesthetic and Chemical Determinants*; SABS Standards Division: Pretoria, South Africa, 2011.
40. Zeliger, H. *Human Toxicology of Chemical Mixtures*, 2nd ed.; William Andrew: Norwich, NY, USA, 2011; p. 54. ISBN 9781437734645.
41. ARGOSS. *Assessing Risk to Groundwater from On-Site Sanitation: Scientific Review and Case Studies*; British Geological Survey Commissioned Report, CR/02/079N, 2002: 4411; British Geological Survey: Nottingham, UK, 2002.
42. Lorentz, S.; Wickham, B.; Still, D. *Investigation into Pollution from On-Site Dry Sanitation Systems*; Report to the Water Research Commission; WRC Report No. 2015/1/15; Water Research Commission: Pretoria, South Africa, 2015; ISBN 978-1-4312-0671-1.
43. Wade, T.J.; Calderon, R.L.; Sams, E.; Beach, M.; Brenner, K.P.; Williams, A.H.; Dufour, A.P. Rapidly measured indicators of recreational water quality are predictive of swimming-associated gastrointestinal illness. *Environ. Health Perspect.* **2006**, *114*, 24–28. [CrossRef]
44. Feachem, R.G.; Bradley, D.J.; Garelick, H.; Mara, D.D. Sanitation and Disease: Health aspects of excreta and wastewater management. In *World Bank Studies in Water Supply and Sanitation*; John Wiley and Sons: Oxford, UK, 1983; Volume 3.
45. Torondel, B. *Sanitation Ventures Literature Review: On-Site Sanitation Waste Characteristics*; London School of Hygiene and Tropical Medicine Publications: London, UK, 2010.
46. Gotaas, H.B.; World Health Organization. *Composting: Sanitary Disposal and Reclamation of Organic Wastes*; World Health Organization Monograph Series; World Health Organization: Geneva, Switzerland, 1956; p. 31. Available online: https://apps.who.int/iris/handle/10665/41665 (accessed on 11 October 2021)ISBN 9241400315.
47. Salam, O.E.A.; Reiad, N.A.; ElShafei, M.M. A study of the removal characteristics of heavy metals from wastewater by low-cost adsorbents. *J. Adv. Res.* **2011**, *2*, 297–303. [CrossRef]
48. Akpor, O.B.; Ohiobor, G.O.; Olaolu, T.D. Heavy Metal Pollutants in Wastewater Effluents: Sources, Effects and Remediation. *Adv. Biosci. Bioeng.* **2014**, *2*, 37–43. [CrossRef]
49. Karvelas, M.; Katsoyiannis, A.; Samara, C. Occurrence and fate of heavy metals in the wastewater treatment process. *Chemosphere* **2003**, *53*, 1201–1210. [CrossRef]
50. Bleuler, M.; Gold, M.; Strande, L.; Schonborn, A. Pyrolysis of Dry Toilet Substrate as a Means of Nutrient Recycling in Agricultural Systems: Potential Risks and Benefits. *Waste Biomass Valorization* **2020**, *7*, 4171–4183. [CrossRef]
51. Latosinska, J.; Kowalik, R.; Gawdzik, J. Risk Assessment of Soil Contamination with Heavy Metals from Municipal Sewage Sludge. *Appl. Sci.* **2021**, *11*, 548. [CrossRef]
52. Wanga, M.; Chena, C.; Jub, Y.; Tsaic, M.; Chena, C.; Donga, C. Distribution and environmental risk assessment of trace metals in sludge from multiple sources in Taiwan. *J. Environ. Sci. Health Part A* **2021**, *56*, 481–491. [CrossRef]
53. Mamera, M.; van Tol, J.J.; Aghoghovwia, M.P.; Kotze, E. Sensitivity and Calibration of the FT-IR Spectroscopy on Concentration of Heavy Metal Ions in River and Borehole Water Sources. *Appl. Sci.* **2020**, *10*, 7785. [CrossRef]
54. Jern, W.N.G. *Industrial Wastewater Treatment*; Imperial College Press: Singapore, 2006.
55. Al-Qahtani, K.M. Water purification using different waste fruit cortexes for the removal of heavy metals. *J. Taibah Univ. Sci.* **2015**, *10*, 700–708.
56. Fernandez, L.G.; Olalla, H.Y. Toxicity and bioaccumulation of lead and cadmium in marine protozoan communities. *Ecotoxicol. Environ. Saf.* **2000**, *47*, 266–276. [CrossRef]
57. Guibaud, G.; Hullebusch, E.V.; Bordas, F. Lead and cadmium biosorption by extracellular polymeric substances (EPS) extracted from activated sludges: pH-sorption edge tests and mathematical equilibrium modelling. *Chemosphere* **2006**, *4*, 1955–1962.
58. Ogoyi, D.O.; Mwita, C.J.; Nguu, E.K.; Shiundu, P.M. Determination of heavy metal content in water, sediment and microalgae from Lake Victoria, East Africa. *Open Environ. Eng. J.* **2011**, *4*, 156–161.
59. Huanga, H.J.; Yuanb, X.Z. The migration and transformation behaviors of heavy metals during the hydrothermal treatment of sewage sludge. *Bioresour. Technol.* **2016**, *200*, 991–998.
60. Koger, S.M.; Schettler, T.; Weiss, B. Environmental Toxicants and Developmental Disabilities: A Challenge for Psychologists. *Am. Psychol.* **2005**, *60*, 243–255. [CrossRef]
61. DWAF. Guidelines for the Utilisation and Disposal of Wastewater Sludge. 2008; Volume 4. Available online: http://www.dwaf.gov.za/Dir_WQM/docs/wastewatersludgeMar08vol4part1.pdf (accessed on 15 June 2021).
62. Zhou, Y.; Gao, B.; Zimmerman, A.R.; Fang, J.; Sun, Y.; Cao, X. Sorption of Heavy Metals on Chitosan-Modified Biochars and its Biological Effects. *Chem. Eng. J.* **2013**, *231*, 512–518.
63. Azizi, S.; Kamika, I.; Tekere, M. Evaluation of Heavy Metal Removal from Wastewater in a Modified Packed Bed Biofilm Reactor. *PLoS ONE* **2016**, *11*, e90972. [CrossRef]
64. Zhang, X.; Wang, X.Q.; Wang, D. Immobilization of Heavy Metals in Sewage Sludge during Land Application Process in China: A Review. *Sustainability* **2016**, *9*, 2020. [CrossRef]

65. Ahmedna, M.; Marshall, W.F.; Husseiny, A.A.; Rao, R.M.; Goktepe, I. The use of nutshell carbons in drinking water filters for removal of trace metals. *Water Resour.* **2004**, *38*, 1064–1068.
66. Piccirillo, C.; Pereira, S.; Marques, A.P.; Pullar, R.; Tobaldi, D.; Pintado, M.E. Bacteria immobilisation on hydroxyapatite surface for heavy metals removal. *J. Environ. Manag.* **2013**, *121*, 87–95. [CrossRef] [PubMed]
67. Viglasova, E.; Galambos, M.; Dankova, Z.; Krivosudsky, L.; Lengauer, C.L.; Hood-Nowotny, R.; Soja, G.; Rompel, A.; Matik, M.; Briancin, J. Production, characterization and adsorption studies of bamboo-based biochar/montmorillonite composite for nitrate removal. *Waste Manag.* **2018**, *79*, 385–394.
68. Zhang, S.; Yang, X.; Liu, L.; Ju, M.; Zheng, K. Adsorption Behavior of Selective Recognition Functionalized Biochar to Cd (II) in Wastewater. *Materials* **2018**, *11*, 299. [CrossRef]
69. Sombroek, W.G.; Kern, D.C.; Rodrigues, T.; Cravo, M.D.S.; Cunha, T.J.; Woods, W.; Glaser, B. Terra preta and Terra Mulata, pre-Colombian kitchen middens and agricultural fields, their sustainability and replication. In Proceedings of the 17th World Congress of Soil Science, Bangkok, Thailand, 14–21 August 2002.
70. Lehmann, J. Bio-energy in the black. *Front. Ecol. Environ.* **2007**, *5*, 381–387. [CrossRef]
71. Parmar, A.; Nema, P.K.; Agarwal, T. Biochar production from agro-food industry residues: A sustainable approach for soil and environmental management. *Curr. Sci.* **2014**, *107*, 1673–1682.
72. Starkenmann, C.; Niclass, Y.; Beaussoubre, P.; Zimmermann, J.; Cayeux, I.; Jean, C.; Chappuis, F.; Wolfgang, F. *Use of Fecal and Sawdust Biochar as a New Perfume Delivery System*; John Wiley & Sons Ltd.: Hoboken, NJ, USA, 2017. [CrossRef]
73. IBI. Standardized Product Definition and Product Testing Guidelines for Biochar That Is Used in Soil. Available online: https://www.biochar-international.org/wp-content/uploads/2018/04/IBI_Biochar_Standards_V2.1_Final.pdf (accessed on 21 March 2021).
74. Sohi, S.; Lopez-Capel, E.; Krull, E.; Bol, R. *Biochar, Climate Change and Soil: A Review to Guide Future Research*; CSIRO: Glen Osmond, Australia, 2009.
75. Patra, J.M.; Panda, S.S.; Dhal, N.K. Biochar as a low-cost adsorbent for heavy metal removal: A review. *Int. J. Res. Biosci.* **2017**, *6*, 1–7, ISSN 2319-2844.
76. Sirajudheen, P.; Karthikeyan, P.; Vigneshwaran, S.; Meenakshi, S. Synthesis and characterization of La(III)supported carboxymethylcellulose-clay composite for toxic dyes removal: Evaluation of adsorption kinetics, isotherms and thermodynamics. *Int. J. Biol. Macromol.* **2020**, *161*, 1117–1126. [CrossRef] [PubMed]
77. Srivatsav, P.; Bhargav, B.S.; Shanmugasundaram, V.; Arun, J.; Gopinath, K.P.; Bhatnagar, A. Biochar as an Eco-Friendly and Economical Adsorbent for the Removal of Colorants (Dyes) from Aqueous Environment: A Review. *Water* **2020**, *12*, 3561. [CrossRef]
78. Morais da Silva, P.M.; Camparotto, N.G.; Grego Lira, K.T.; Franco Picone, C.S.; Prediger, P. Adsorptive removal of basic dye onto sustainable chitosan beads: Equilibrium, kinetics, stability, continuous-mode adsorption and mechanism. *Sustain. Chem. Pharm.* **2020**, *18*, 100318. [CrossRef]
79. Inyang, M.I.; Gao, B.; Yao, Y.; Xue, Y.; Zimmerman, A.; Mosa, A.; Pullammanappallil, P.; Sik Ok, Y.; Cao, X. A review of biochar as a low-cost adsorbent for aqueous heavy metal removal, Critical Reviews. *Environ. Sci. Technol.* **2016**, *46*, 406–433. [CrossRef]
80. Duku, M.H.; Gu, S.; Hagan, E.B. Biochar production potential in Ghana—A review. *Renew. Sustain. Energy Rev.* **2011**, *15*, 3539–3551. [CrossRef]
81. Sohi, S.P.; Krull, E.; Lopez-Capel, E.; Bol, R. A Review of Biochar and Its Use and Function in Soil. In *Advances in Agronomy*; Sparks, D.L., Ed.; Academic Press: Burlington, ON, Canada, 2010; Volume 105, pp. 47–82. ISBN 978-0-12-381023-6.
82. Purakayastha, T.J.; Kumari, S.; Pathak, H. Characterisation, stability, and microbial effects of four biochars produced from crop residues. *Geoderma* **2015**, *239*, 293–303. [CrossRef]
83. Leng, L.; Xiong, Q.; Yang, L.; Li, H.; Zhou, Y.; Zhang, W.; Jiang, S.; Li, H.; Huang, H. An overview on engineering the surface area and porosity of biochar. *Sci. Total Environ.* **2021**, *763*, 144204. [CrossRef] [PubMed]
84. Sika, M.; Hardie, A. Effect of pine wood biochar on ammonium nitrate leaching and availability in a South African sandy soil. *Eur. J. Soil Sci.* **2014**, *65*, 113–119. [CrossRef]
85. Jeffery, S.; Verheijen, F.G.A.; van der Velde, M.; Bastos, A.C. A quantitative review of the effects of biochar application to soils on crop productivity using meta-analysis. *Agric. Ecosyst. Environ.* **2011**, *144*, 175–187. [CrossRef]
86. Gwenzi, W.; Chaukura, N.; Mukome, F.N.D.; Machado, S.; Nyamasoka, B. Biochar production and applications in sub-Saharan Africa: Opportunities, constraints, risks and uncertainties. *J. Environ. Manag.* **2015**, *150*, 250–261. [CrossRef] [PubMed]
87. Xu, G.; Lv, Y.; Sun, J.; Shao, H.; Wei, L. Recent advances in biochar applications in agricultural soils: Benefits and environmental implications. *Clean–Soil Air Water* **2015**, *40*, 1093–1098. [CrossRef]
88. Schulz, H.; Glaser, B. Effects of biochar compared to organic and inorganic fertilizers on soil quality and plant growth in a greenhouse experiment. *J. Plant Nutr. Soil Sci.* **2012**, *175*, 410–422. [CrossRef]
89. Schnell, R.W.; Vietor, D.M.; Provin, T.L.; Munster, C.L.; Capareda, S. Capacity of Biochar Application to Maintain Energy Crop Productivity: Soil Chemistry, Sorghum Growth, and Runoff Water Quality Effects. *J. Environ. Qual.* **2012**, *41*, 1044–1051. [CrossRef]
90. Zhai, L.M.; Caiji, Z.M.; Liu, J.; Wang, H.Y.; Ren, T.Z.; Gai, X.P.; Xi, B.; Liu, H.B. Short-term effects of maize residue biochar on phosphorus availability in two soils with different phosphorus sorption capacities. *Biol. Fertil. Soils* **2015**, *51*, 113–122. [CrossRef]

91. Novak, J.M.; Lima, I.; Xing, B.; Gaskin, J.W.; Steiner, C.; Das, K.C.; Ahmedna, M.; Rehrah, D.; Watts, D.W.; Busscher, W.J.; et al. Characterization of Designer Biochar Produced at Different Temperatures and Their Effects on a Loamy Sand. *Ann. Environ. Sci.* **2009**, *3*, 195–206.
92. Beck, D.A.; Johnson, G.R.; Spolek, G.A. Amending green roof soil with biochar to affect runoff water quantity and quality. *Environ. Pollut.* **2011**, *159*, 2111–2118. [PubMed]
93. Ulyett, J.; Sakrabani, R.; Kibblewhite, M.; Hann, M. Impact of Biochar Addition on Water Retention, Nitrification, and Carbon Dioxide Evolution of Two Sandy Loam Soils. *Eur. J. Soil Sci.* **2014**, *65*, 96–104. [CrossRef]
94. Laird, D. The Charcoal Vision: A Win-Win-Win Scenario for Simultaneously Producing Bioenergy, Permanently Sequestering Carbon, while Improving Soil and Water Quality. *Agron. J.* **2008**, *100*, 178–181. [CrossRef]
95. Park, J.H.; Choppala, G.K.; Bolan, N.S.; Chung, J.W.; Chuasavathi, T. Biochar Reduces the Bioavailability and Phytotoxicity of Heavy Metals. *Plant Soil* **2011**, *348*, 439–451.
96. Zhang, H.; Nordin, N.A.; Olson, M.S. Evaluating the effects of variable water chemistry on bacterial transport during infiltration. *J. Contam. Hydrol.* **2013**, *150*, 54–64. [CrossRef]
97. Lehmann, J.; Gaunt, J.; Rondon, M. Bio-char Sequestration in Terrestrial Ecosystems—A Review. *Mitig. Adapt. Strateg. Glob. Chang.* **2006**, *11*, 403–427. [CrossRef]
98. Lentz, R.D.; Ippolito, J.A.; Spokas, K.A. Biochar and Manure Effects on Net Nitrogen Mineralization and Greenhouse Gas Emissions from Calcareous Soil under Corn. *Soil Sci. Soc. Am. J.* **2014**, *78*, 1641–1655. [CrossRef]
99. Downie, A.; van Zwieten, L. Biochar: A Coproduct to Bioenergy from Slow-Pyrolysis Technology. In *Advanced Biofuels and Bioproducts*; Springer: New York, NY, USA, 2013; pp. 97–117.
100. Atkinson, C.J.; Fitzgerald, J.D.; Hipps, N.A. Potential mechanisms for achieving agricultural benefits from biochar application to temperate soils: A review. *Plant Soil* **2010**, *337*, 1–18. [CrossRef]
101. Chan, K.Y.; Van Zwieten, L.; Meszaros, I.; Downie, A.; Joseph, S. Using poultry litter biochars as soil amendments. *Soil Resour.* **2008**, *46*, 437–444. [CrossRef]
102. Glaser, B.; Lehmann, J.; Zech, W. Ameliorating physical and chemical properties of highly weathered soils in the tropics with charcoal—A review. *Biol. Fertil. Soils* **2002**, *35*, 219–230. [CrossRef]
103. Downie, A.; Crosky, A.; Munroe, P. Physical properties of biochar. In *Biochar for Environmental Management: Science and Technology*; Lehmann, J., Joseph, S., Eds.; Earthscan: London, UK, 2009; pp. 13–32.
104. Brady, N.C.; Weil, R.R. *An Introduction to the Nature and Properties of Soils*, 14th ed.; Prentice Hall: Upper Saddle River, NJ, USA, 2008.
105. Lehmann, J.; Rondon, M. Bio-Char Soil Management on Highly Weathered Soils in the Humid Tropics. In *Biological Approaches to Sustainable Soil Systems*; Uphoff, N., Ed.; Taylor & Francis Group: Boca Raton, FL, USA, 2006.
106. Huggins, T.M.; Haeger, A.; Biffinger, J.C.; Zhiyong, J.R. Granular biochar compared with activated carbon for wastewater treatment and resource recovery. *Water Res.* **2016**, *94*, 225–232. [CrossRef] [PubMed]
107. Kumar, S.; Loganathan, V.A.; Gupta, R.B.; Barnett, M.O. An Assessment of U (VI) removal from groundwater using biochar produced from hydrothermal carbonization. *J. Environ. Manag.* **2011**, *92*, 2504–2512.
108. Wang, S.G.; Zimmerman, A.; Li, Y.; Ma, L.; Harris, W.; Migliaccio, K. Removal of arsenic by magnetic biochar prepared from pinewood and natural hematite. *Bioresour. Technol.* **2014**, *175*, 391–395. [CrossRef] [PubMed]
109. Field, J.L.; Keske, C.M.H.; Birch, G.L.; Defoort, M.W.; Cotrufo, M.F. Distributed biochar and bioenergy coproduction: A regionally specific case study of environmental benefits and economic impacts. *Glob. Chang. Biol. Bioenergy* **2013**, *5*, 177–191.
110. Chen, X.; Chen, G.; Chen, L.; Chen, Y.; Lehmann, J.; McBride, M.B.; Hay, A.G. Adsorption of Copper and Zinc by Biochars Produced from Pyrolysis of Hardwood and Corn Straw in Aqueous Solution. *Bioresour. Technol.* **2011**, *102*, 8877–8884. [CrossRef]
111. Krueger, B.C.; Fowler, G.D.; Templeton, M.R.; Moya, B. Resource recovery and biochar characteristics from full-scale faecal sludge treatment and co-treatment with agricultural waste. *Water Res.* **2020**, *169*, 115253. [CrossRef]
112. Bolster, C.H.; Walker, S.L.; Cook, K.L. Comparison of *Escherichia coli* and *Campylobacter jejuni* Transport in Saturated Porous Media. *J. Environ. Qual.* **2006**, *35*, 1018–1025. [CrossRef] [PubMed]
113. Harvey, R.W.; Metge, D.W.; Mohanram, A.; Gao, X.; Chorover, J. Differential Effects of Dissolved Organic Carbon upon Re-Entrainment and Surface Properties of Groundwater Bacteria and Bacteria-Sized Micro-Spheres during Transport through a Contaminated, Sandy Aquifer. *Environ. Sci. Technol.* **2011**, *45*, 3252–3259. [CrossRef]
114. Kim, H.N.; Bradford, S.A.; Walker, S.L. *Escherichia coli* O157:H7 Transport in Saturated Porous Media: Role of Solution Chemistry and Surface Macromolecules. *Environ. Sci. Technol.* **2009**, *43*, 4340–4347. [CrossRef] [PubMed]
115. Kranner, B.P.; Afrooz, A.R.M.N.; Fitzgerald, N.J.M.; Boehm, A.B. Fecal indicator bacteria and virus removal in stormwater biofilters: Effects of biochar, media saturation, and field conditioning. *PLoS ONE* **2019**, *14*, 222–719.
116. Chen, X.; He, H.Z.; Chen, G.K.; Li, H.S. Effects of biochar and crop straws on the bioavailability of cadmium in contaminated soil. *Sci. Rep.* **2020**, *10*, 9528. [CrossRef] [PubMed]

Article

Removal of Emerging Contaminants as Diclofenac and Caffeine Using Activated Carbon Obtained from Argan Fruit Shells

Badr Bouhcain [1,*], Daniela Carrillo-Peña [2], Fouad El Mansouri [3], Yassine Ez Zoubi [1], Raúl Mateos [2], Antonio Morán [2,*], José María Quiroga [4] and Mohammed Hassani Zerrouk [1,*]

[1] Environmental Technologies, Biotechnology and Valorisation of Bio-Resources Team, TEBVB, FSTH, Abdelmalek Essaadi University, Tetouan 93020, Morocco; y.ezzoubi@uae.ac.ma
[2] Chemical and Environmental Bioprocess Engineering Group, Natural Resources Institute, University of León, 24071 León, Spain; dcarp@unileon.es (D.C.-P.); rmatg@unileon.es (R.M.)
[3] Laboratory of Chemical Engineering and Valorisation of Resources, Department of Chemistry, Faculty of Sciences and Technology, Abdelmalek Essaâdi University, Tangier 416, Morocco; fouad.elmansouri@etu.uae.ac.ma
[4] Department of Chemical Engineering, Food Technologies and Environmental Technologies, Faculty of Marine and Environmental Sciences, University of Cádiz, 11510 Puerto Real Cádiz, Spain; josemaria.quiroga@uca.es
* Correspondence: badr.bouhcain@etu.uae.ac.ma (B.B.); amorp@unileon.es (A.M.); m.hassani@uae.ac.ma (M.H.Z.)

Abstract: Activated carbons from argan nutshells were prepared by chemical activation using phosphoric acid H_3PO_4. This material was characterized by thermogravimetric analysis, infrared spectrometry, and the Brunauer–Emmett–Teller method. The adsorption of two emerging compounds, a stimulant caffeine and an anti-inflammatory drug diclofenac, from distilled water through batch and dynamic tests was investigated. Batch mode experiments were conducted to assess the capacity of adsorption of caffeine and diclofenac from an aqueous solution using the carbon above. Adsorption tests showed that the equilibrium time is 60 and 90 min for diclofenac and caffeine, respectively. The adsorption of diclofenac and caffeine on activated carbon from argan nutshells is described by a pseudo-second-order kinetic model. The highest adsorption capacity determined by the mathematical model of Langmuir is about 126 mg/g for diclofenac and 210 mg/g for caffeine. The thermodynamic parameters attached to the studied absorbent/adsorbate system indicate that the adsorption process is spontaneous and exothermic for diclofenac and endothermic for caffeine.

Keywords: activated carbon; adsorption; caffeine; diclofenac; argan nutshells; emerging contaminants

Citation: Bouhcain, B.; Carrillo-Peña, D.; El Mansouri, F.; Ez Zoubi, Y.; Mateos, R.; Morán, A.; Quiroga, J.M.; Zerrouk, M.H. Removal of Emerging Contaminants as Diclofenac and Caffeine Using Activated Carbon Obtained from Argan Fruit Shells. *Appl. Sci.* **2022**, *12*, 2922. https://doi.org/10.3390/app12062922

Academic Editors: Amanda Laca Pérez and Yolanda Patiño

Received: 21 January 2022
Accepted: 11 February 2022
Published: 12 March 2022

Publisher's Note: MDPI stays neutral with regard to jurisdictional claims in published maps and institutional affiliations.

Copyright: © 2022 by the authors. Licensee MDPI, Basel, Switzerland. This article is an open access article distributed under the terms and conditions of the Creative Commons Attribution (CC BY) license (https://creativecommons.org/licenses/by/4.0/).

1. Introduction

With the continuous increase of human demand for the environment, many pollutants with low content in the environment but with great harm have gradually attracted people's attention, such as anti-inflammatories, antibiotics, etc., which are called emerging contaminants (ECs). ECs are a group of chemical pollutants that have potential threats to human health and the ecological environment. They are very complex organic matters and generally exist in water. ECs usually comes from medicines, personal care products, endocrine-disrupting chemicals, antibiotics, persistent organic pollutants, disinfection by-products, and other industrial chemicals [1]. These ECs persist in the environment and last for a long time. Previous studies have found more than 30 ECs in untreated wastewater, treated wastewater, urban rainwater, agricultural rainwater, and fresh water. Among them, artificial sweeteners, pharmaceuticals, and personal care products were detected in various water samples [2]. ECs are constantly circulating, migrating, and transforming in environmental media. Although the concentration of these ECs in water is relatively low, they may have potential impacts on the environment and human health through the food chain after being accumulated by organisms [3]. Therefore, how to effectively remove ECs in water has received widespread attention.

This is the case of pharmaceutical drugs, and this is the main theme of this work. For example, in the case of painkillers and anti-inflammatory drugs such as diclofenac (Dic), their ecological toxicity and the removal capacity of conventional wastewater treatment plants are worrying. This drug has been frequently seen in wastewater and surface water at concentrations up to 2 µg/L [4], and its chronic effects need to be analyzed. Another example is caffeine (Caf), a psychostimulant and analeptic that is largely consumed by the human population and expelled basically in urine. It is frequently found in surface waters, and indeed at low concentrations caffeine can negatively affect the metabolism of fish, amphibians, and reptiles [5–8].

Diclofenac and caffeine have been removed by using different types of adsorbents which are listed in Table 1, along with percentage removal.

Table 1. Activated carbon performance from different agricultural waste toward caffeine and diclofenac.

Absorbate	Absorbent	Initial Concentration	Absorbent Dosage	Time	Adsorption Capacity	References
Caffeine	Peach stones	100 mg	0.12 g	2 h	126 mg/g	[9]
	Acacia mangium wood	100 mg	3 g	61 min	30.9 mg/g	[10]
	Date stone	100 mg	1 g	80 min	28 mg/g	[11]
	Macrophytes	150 mg	1 g	1 h	117.8 mg/g	[12]
	Açaí seed	300 mg	1 g	3 h	176.8 mg/g	[13]
	Pineapple Plant leaves	500 mg	1 g	4 h	152 mg/g	[14]
	Sargassum	20 mg	0.6 g	90 min	221.6 mg/g	[15]
	Pine Wood	120 mg	0.3 g	5 h	362 mg/g	[16]
	Coffee waste	25 mg	0.1 g	30 min	274.2 mg/g	[17]
	Elaeis guineensis	20 mg	0.2 g	5 h	13.5 mg/g	[18]
	Eragrostis Plana Nees leaves	200 mg	0.07 g	1 h	235.5 mg/g	[19]
	Argan nutshells	100 mg	1 g	90 min	210.65 mg/g	This work
Diclofenac	Sycamore ball	150 mg	0.2 g	2 h	178.9 mg/g	[20]
	Pine tree	100 mg	0.8 g	2 h	54.67 mg/g	[21]
	Sugar cane bagasse	50 mg	0.4 g	15 min	315 mg/g	[22]
	Cocoa shell	150 mg	1 g	223 min	63.47 mg/g	[23]
	Tea waste	30 mg	0.3 g	8 h	62.5 mg/g	[24]
	Potato peel waste	50 mg	0.4 g	24 h	68.5 mg/g	[25]
	Olive-waste cakes	50 mg	0.1 g	26 h	56.2 mg/g	[26]
	Pine sawdust-Onopordum acanthium	100 mg	2.4 g	1 h	263.7 mg/g	[27]
	Coconut shell	200 mg	0.5 g	24 h	103 mg/g	[28]
	Peach stones	100 mg	0.12 g	2 h	200 mg/g	[9]
	Orange peels	0.5 mM	0.5 g	24 h	52.2 mg/g	[29]
	Argan nutshells	100 mg	1 g	90 min	126.16 mg/g	Present work

At present, the methods generally used to remove ECs from water basically include the microbial method [30], electrochemical method [31], adsorption method, membrane process, and chemical oxidation process [32]. Among them, adsorption is extensively accepted because of its advantages such as low cost, high efficiency, and wide processing range. Generally used adsorbents include activated carbon [33–35]. The mechanisms of adsorption are usually non-specific which could be employed to eliminate or reduce a large variety of contaminants [36]. Adsorption is a widely acknowledged surface phenomenon which is also a method for equilibrium separation as an effective process for removal of pollutants from the wastewater [37–40]. Adsorption was observed to be advantageous over other wastewater treatment methods in terms of initial price, simple design, ease of use and non-sensitivity to harmful substances. Adsorption is therefore not allowing hazardous chemicals to form [41–43]. Presently, activated carbon is the most widely used adsorbent. It is substantially used to eliminate complex pollutants from wastewater, like dyes and heavy metals [44].

Activated carbon (AC) is a long-known adsorbent distinguished by, among other things, its large specific surface area, porous structure, and thermostability [45]. Activated carbon might be prepared from any solid material containing a high proportion of carbon often by carbonization followed by physical or chemical activation. However, a process combining both steps can be applied [46]. Carbonization is essentially aimed at enriching the material in carbon and creating the first pores, while activation aims at developing a porous structure [47]. Good-quality activated carbons are prepared by plant biomass using orthophosphoric acid H_3PO_4 as a chemical activating agent [48].

The AC resulting from these treatments acquires an adsorbing [49] and catalyzing capacity [50], which is highly sought after in several fields [46]: the pharmaceutical, food, and automotive industries. AC is widely used in water purification. It allows for the removal of organic (e.g., pesticides) and inorganic (e.g., heavy metals such as Pb) materials [51].

During the last decade, the ability of agricultural by-products to give ACs with a high adsorption capacity and very advantageous physicochemical properties including, among others, a low ash content, has not ceased to attract the attention of researchers [46]. Numerous works have been undertaken on the plant material of various origins: corn straw [52], olive pits [53], sunflower seed shells [54], sugarcane bagasse [55], almond shells [56], peach pits [57], grape seeds [58], apricot kernels [56], cherry pits [58], peanut shells [59], walnut shells [60], rice hulls [61], corn hulls [62], and barley seeds [63].

In Morocco, the agricultural activity attached to the production of argan fruit (Argania Spinosa) for oil extraction is rapidly emerging because of developing interest regarding its usages for culinary and cosmetic purposes worldwide. So far, the increased popularity of argan oil has prompted an annual production up to 4000 tons by Morocco, which leaves behind about 80,000 tons of hard shells [64]. The latter is currently considered as an agriculture by-product without any significant economic value and is mainly used by the local population as a domestic combustible [65]. Even more interesting, argan shells are well known by their rich lignocellulosic content [66], with high potential for use as raw material to produce activated carbons.

Indeed, we previously reported successful production of nanoporous activated carbon made from argan shells using optimal preparation conditions following an empirical approach [67]. The purpose of the present work was to initially obtain an activated carbon by chemical activation of argan fruit shells, then to investigate its capacity of adsorption on caffeine and diclofenac. This property is determined by the depollution of various industrial effluents.

2. Materials and Methods

2.1. Materials

The argan nutshells studied were collected in September 2020 in the rural area of the region of Tafraout (29°43′11.1″ N 8°58′51.7″ W), southeast Morocco. These are waste fruits from Argan trees that grow spontaneously but do not benefit from any valuation. Figure 1 shows the initial samples of argan fruit.

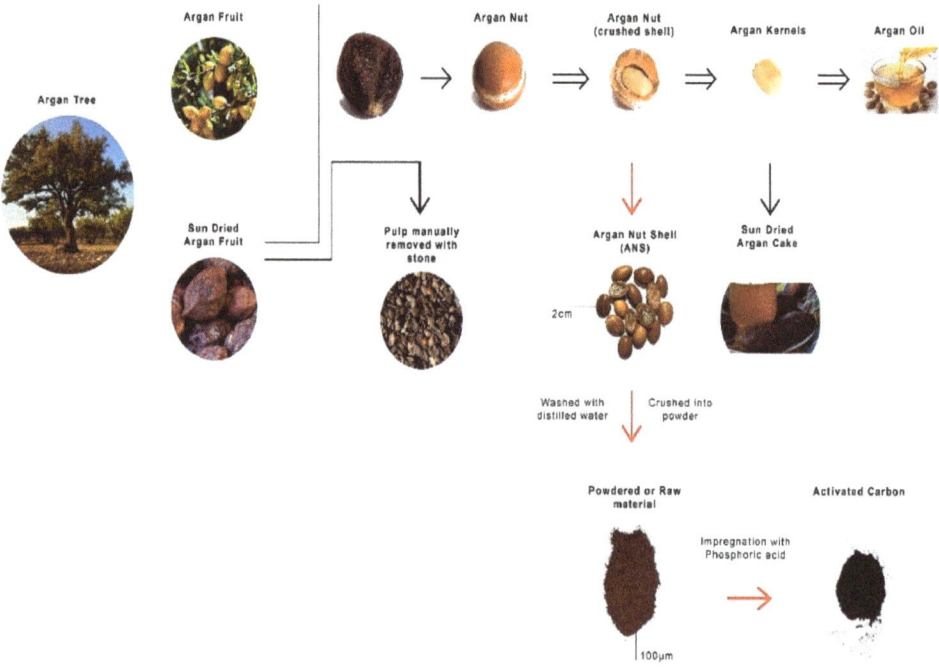

Figure 1. Initial samples of argan fruit and the process of preparation of activated carbon.

Caffeine anhydrous 98.5% (Caf) was supplied by PanReac AppliChem and diclofenac sodium 98% (Dic) was purchased from Acros Organics. Their chemical structure and other properties are shown in Table 2.

Table 2. Properties and chemical structures of the contaminants studied.

Emerging Contaminant	Mass Molar (g/Mol)	PKa	Size (nm)	Chemical Structure
Caffeine	194.2	0.82	0.98–0.87	
Diclofenac	318.1	4.15	0.97–0.96	

2.2. Preparation of the Activated Carbon

The preparation process of the AC (Figure 1) has two steps: carbonization and activation. In the first step, argan was crushed and sieved to get a particle size ranging between 80 and 100 μm. Then it was washed with distilled water and dried at 60 °C for 24 h. In the second step, the chemical impregnation was done in a round-bottom flask reactor, where 60 g of argan reacted with an H_3PO_4 solution 85% (1:3) for 24 h at an ambient temperature of 25 °C. After impregnation, the solid was filtered under a vacuum to remove the excess phosphoric acid. Then the argan's powder was pyrolyzed at 575 °C for 90 min in a muffle furnace (PR Series Hobersal). Furthermore, the carbon was completely washed with ultra-pure water in order to remove the remaining phosphoric acid until reaching a pH of (6.5). Finally, the samples were dried in the oven at 105 °C for 24 h.

2.3. Characterization of the Activated Carbon

Thermogravimetric analysis (TGA) was used to estimate the temperature distribution at which the nutshell of argan responds under a latent climate. Thermal analyses were done with STD 2960 TA and SDT Q600 instruments under a nitrogen flow of 100 mL/min. The temperature ramp of 10 °C/min from room temperature to 800 °C was utilized during the analyses.

The surface functionalities were investigated with FT-IR spectroscopy. A Thermo IS5 Nicolet (USA) spectrophotometer was used for obtaining FT-IR spectra and acquired from 400 to 4000 cm^{-1} at room temperature (16 scans and spectral resolution of 4 cm^{-1}); the peak positions were determined using Origin software (Version 2021b). Origin Lab Corporation, Northampton, MA, USA.

The textural properties of activated carbons were determined from nitrogen adsorption at 196 °C using a Micrometrics ASAP 2420 (V2.09). Specific surface areas (S_{BET}) were determined by applying the Brunauer–Emmett–Teller (BET) equation to the isotherms. Additionally, the total pore volume (V_{TP}), which corresponds to the N_2 volume adsorbed at a relative pressure (P/P°) of 0.95, was calculated. The volume of the micropores ($V_{\mu P}$) and external surface area (S_{EXT}) were determined using the t-plot method. The external volume (V_{EXT}) was calculated using the difference between V_{TP} and $V_{\mu P}$. The average pore diameter (D_{AP}) was calculated using the $4V_{TP}/S_{BET}$ ratio.

2.4. Adsorptions Experiments

Adsorption is a surface phenomenon in which only the adsorbent surface is concerned, and adsorbate should not penetrate inside the structure of the adsorbent. Figure 2 depicts the adsorption process.

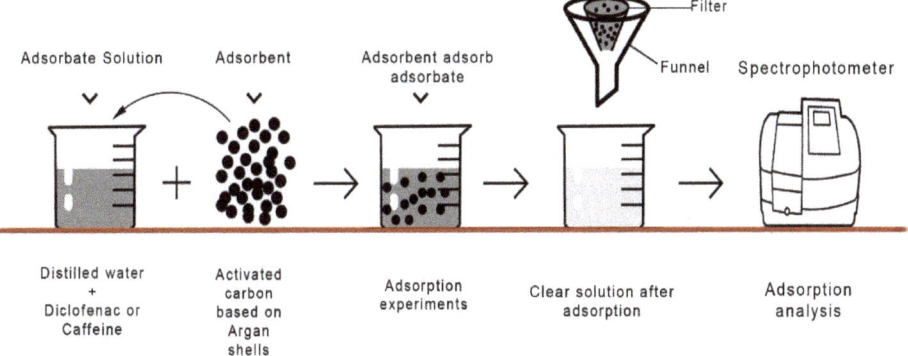

Figure 2. Adsorption process.

Batch adsorption experiments were performed on IKA Magnetic stirrers (RO 15) with a Digiterm 100 microprocessor-controlled digital immersion thermostat and thermostatic circulating bath. In addition, a magnetic bar was added to stir the solution and a weight circle was added to avoid floating. In order to obtain the adsorption equilibrium time, the evolution of the adsorbate concentration was studied by adding 1 g of activated carbon to adsorbate solutions (Co = 100 mg/L) in 50 mL-flask. The experiments were carried out at controlled shaking (200 rpm) and temperature (30 °C) until reaching equilibrium.

The amount of adsorbed compound at equilibrium time, which represents the adsorption capacity, Qe (mg/g), and Qt is the amount of adsorbed compound at random time t, can be determined by the next expressions:

$$Qe = \frac{(Co - Ce).V}{W} \qquad (1)$$

$$Qt = \frac{(Co - Ct).V}{W} \qquad (2)$$

where Co, Ct, and Ce (mg/L) are the absorbate concentrations at beginning, at time t, and at equilibrium, respectively; V is the volume of solution (L) and W is the weight of adsorbent (g).

When the adsorption equilibrium was reached, the adsorbent was removed from the solution by filtration with syringe filters (0.45 µm) and the residual adsorbate concentration was analyzed by a VWR UV-1600PC spectrophotometer. All experiments were carried out at a natural pH of the adsorbate solution at the maximum absorbance wavelength (λmax) of 300 nm for diclofenac and 290 nm for caffeine. The obtained data were adjusted to the Langmuir and Freundlich isotherm models. The kinetic models of pseudo-first order and pseudo-second order were evaluated.

2.4.1. Adsorption Kinetic

The mechanism through which adsorbate particles bind to the absorbent surface is adsorption. Through column or section configuration, the adsorption process is achieved. Kinetic studies are a curve (or line) which characterizes the speed of persistence or transfer of a solution at a given adsorbent dosage, temperature, and pH with an aqueous atmosphere to phase boundaries. Two major processes occur in adsorption: physical adsorption and chemical adsorption. Physical adsorption is due to poor attraction forces (van der Waals), whereas chemical adsorption requires the creation of a tight bond that facilitates the activation of atoms in between the solvent and the substrate [68,69].

Pseudo-first-order kinetic model is a simple kinetic model which describes the process of adsorption and is the pseudo-first-order equation suggested by Lagergren [70,71].

$$Qt = Qe[1 - \exp(-K_1.t)] \qquad (3)$$

where Qe (mg/g) is the amount of the contaminants adsorbed at equilibrium, Qt (mg/g) is the amount of Dic and Caf adsorbed at time t (min), k_1 (L/min) is the rate constant of the pseudo-first-order adsorption.

Pseudo-second-order kinetic model is the kinetic equation that was developed for the adsorption process [72]. The equations are given below:

$$Qt = Qe\left(\frac{Qe.K_2.t}{1 + Qe.K_2.t}\right) \qquad (4)$$

where Qe (mg/g) is the amount of the contaminants adsorbed at equilibrium, Qt (mg/g) is the amount of Dic and Caf adsorbed at time t (min), k_2 (g/mg. min) is the rate constant of the second-order adsorption.

2.4.2. Adsorption Isotherms

Any adsorption system's isotherm is an equation which relates to the amount of adsorbate on the adsorbent surface and the adsorbent's concentration or partial pressure at constant temperature [73]. The most used adsorption isotherms model contaminants for removal are the Langmuir isotherm, Freundlich isotherm, Temkin isotherm, and BET (Brunauer–Emmett–Teller) isotherm which are used to gain extensive knowledge on the relationships between the adsorbent surface and the adsorbate [74,75]. Two classic isotherm equations, namely Langmuir and Freundlich, were selected in this study to determine the isotherm parameters.

Langmuir adsorption is made up of four assumptions. The adsorbent's surface is homogenous, implying that practically all binding sites are equal. Adsorbed molecules do not encounter each other. The method of adsorption is similar in all situations, where a monolayer is always assumed to be formed. It has been developed to clarify gas–solid adsorption where monolayer adsorption is directly proportionate to the fraction of the adsorbent surface, which is opened, while desorption is proportional to the portion of the adsorbent surface covered. The Langmuir isotherm is given as [76,77].

$$Qe = \frac{Qm.Kl.Ce}{1 + Kl.Ce} \quad (5)$$

where Ce is adsorbate's concentration at equilibrium (mg/L), Qm is quantity of molecules adsorbed on the adsorbent's surface at any time (mg/g), and Kl is the Langmuir constant (L/mg). When Ce/Qe is plotted against Ce, a straight line with a slope of 1/Qm and an intercept of 1/Kl Qm is obtained.

Freundlich isotherm maintains multi-layer as well as heterogeneous molecular adsorption and gives an interpretation that describes the heterogeneity of the surface, and furthermore, the exponential function of the active site and their energy [78,79]. The mathematical expression of Freundlich isotherm is:

$$Qe = Kf.Ce^{1/n} \quad (6)$$

where Kf is Freundlich constant or adsorption capacity (L/mg), n represents the extent of heterogeneity in the surface and furthermore characterizes how the adsorbate is distributed on the adsorbent surface. In addition, the exponent (1/n) indicates the absorbent system's favorability and efficiency. As ln (Qe) is plotted against ln (Ce), a straight line with a slope of 1/n and an intercept of ln (Kf) emerges.

2.4.3. Thermodynamic Study of the Adsorption

Thermodynamic parameters, specifically free energy, enthalpy, and entropy changes of adsorption, were assessed utilizing Vant Hoff's equation expressed as follows [80]:

$$\ln Kl = -\frac{\Delta G^\circ}{RT} = \frac{\Delta S^\circ}{R} - \frac{\Delta H^\circ}{RT'} \quad (7)$$

where ΔG° is free energy of adsorption (J/mol), ΔH° is change in enthalpy (J/mol), and ΔS° is change in entropy (J/mol/k).

Energy and entropy factors must be considered for every adsorption process in terms of deciding whether the process has taken effect spontaneously. Thermodynamic variable measurements are the exact metrics for the functional operation of the method [81,82]. Consequently, if the adsorption rate temperature progresses, (ΔH°) > 0, the mechanism is endothermic, or (ΔH°) < 0, the mechanism is exothermic [83,84].

3. Results and Discussion

3.1. Characterization of the Activated Carbon

As displayed in Figure 3, the profile of thermogravimetric analysis (TGA) obtained with argan nutshells clearly shows weight loss occurring as function of temperature increase. This profile is likewise of interest regarding the carbonization temperature range needed for the activated carbon production. In concurrence with the writing [85], the first weight loss of 5.9% is credited to the released of moisture content and volatile matter at a temperature range between 20 °C and 100 °C. The second decomposition stage of the profile shows a weight loss of 61.9% at a temperature range of 240 °C to 370 °C and is due to the decomposition of hemicellulose and cellulose. The final stage of the profile exhibited weight loss of 12.6% and is credited to the decomposition of lignin at a temperature above 370 °C. Stabilization of the material was seen near 600 °C and explains the consideration of this temperature for carbonization.

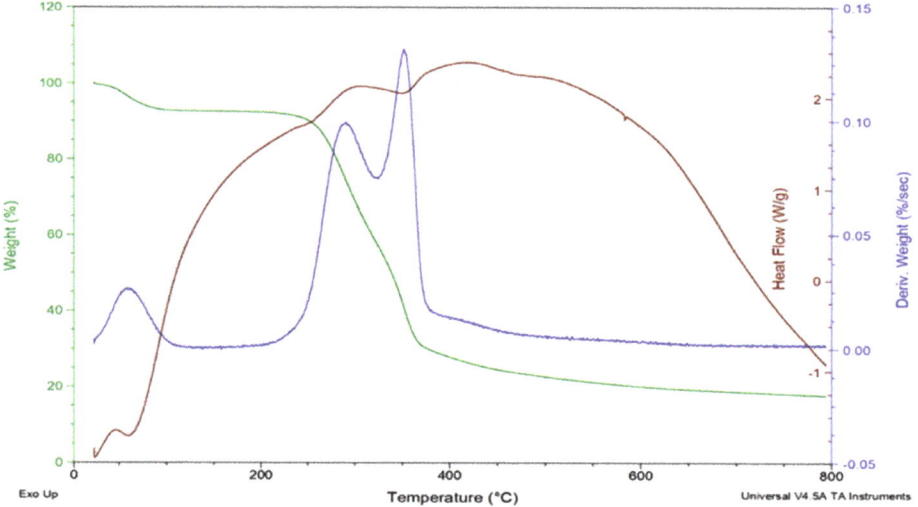

Figure 3. TGA/DSC curve of the argan shells under nitrogen atmosphere.

The FT-IR spectrum of AC, displayed in Figure 4, shows characteristic vibration bands of carbonaceous materials [86]. The figure of spectrum FTIR shows the presence of aromatic amines between 1500 to 1600 cm^{-1}, C-O bonds of Ester between 1210 to 1260 cm^{-1}, the isopropyl group $(CH_3)_2CH$- bonds between 990 to 1050 cm^{-1}, and C-N bonds of the nitrile derivatives at 834 cm^{-1}.

The textural properties of the AC were measured by nitrogen physisorption at 77K. It was evident that AC presented the type II physisorption isotherm (Figure 5) according to IUPAC classification [87], which is characteristic for the microporous materials. The results show that the phosphoric acid obtained the highest specific surface area, highest pore volume, and narrow pore size distribution (Table 3). These properties offer a good potential for the prepared activated carbons to be used as efficient adsorbents.

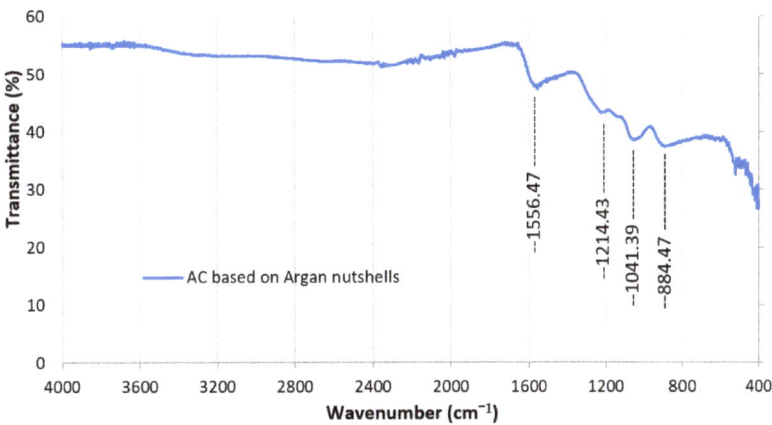

Figure 4. FTIR spectra of activated carbon from argan nutshells.

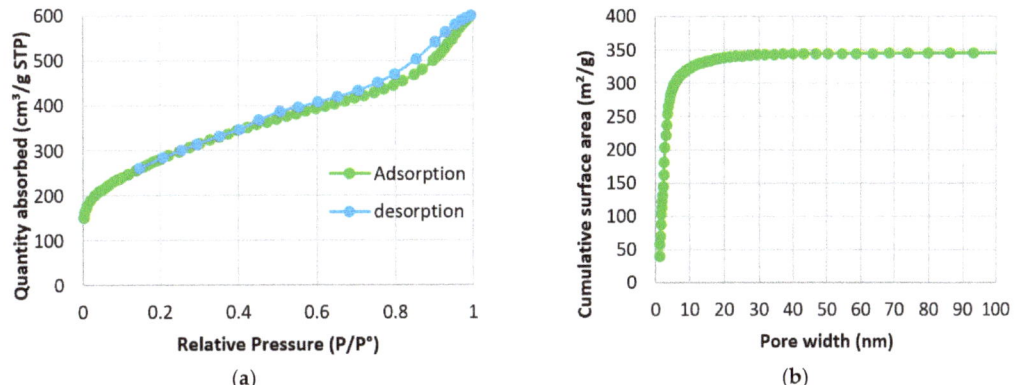

Figure 5. (a) Nitrogen adsorption/desorption isotherms; (b) pore size distribution with insert in the region of pore diameter between 0 and 100 nm for AC based on argan nutshells.

Table 3. Textural properties of the activated carbons obtained from argan.

Absorbent	BET Surface Area (m²/g)	Dubinin-Radushkevich Surface Area (m²/g)	Dubinin-Astakhov Surface Area (m²/g)	Total Pore Volume (cm³/g)	Average Pore Diameter (nm)
AC obtained from argan	1007.76	1063.70	1042.50	0.85	3.38

3.2. Adsorption of Emergent Contaminants

3.2.1. Effect of Contact Time and Adsorption Kinetic

We studied the adsorption efficiency of the two emerging contaminants while modifying the contact time 15, 30, 60, 90, 120 and 150 min. Samples for analysis were taken at regular time intervals to determine the percent removal of contaminants. The results obtained are shown in Figure 6.

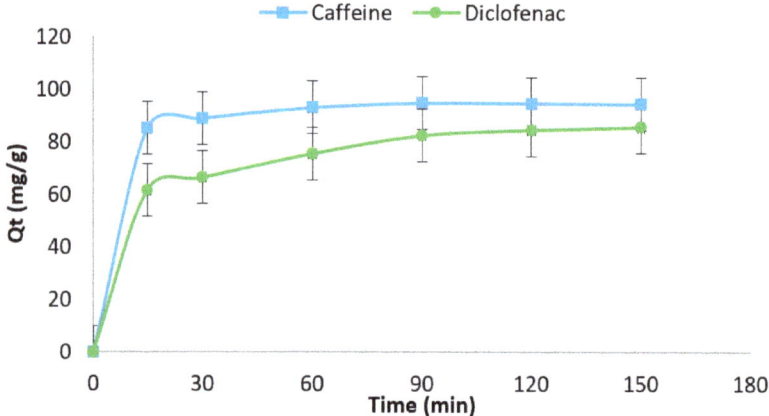

Figure 6. Adsorption of Dic and Caf onto AC based on argan nutshells at different temperatures (C_0 = 100 mg/L; m = 1 g; T = 30 °C; agitation speed = 200 rpm).

Adsorption kinetics of Caf and Dic showed that they were adsorbed rapidly at the investigated conditions, with equilibration already achieved at 90 min of contact for Dic and 60 min for Caf (Figure 6).

The absorbance quantity of Dic and Caf at the equilibrium was 82% and 92%, with experimental uptake capacities of 82.60 mg/g and 93.09 mg/g, respectively.

This information indicates that all adsorption data obtained after these times can be considered as obtained under equilibrium conditions. It is necessary to identify the step that governs the overall removal rate in the above adsorption process. The pseudo-first-order and pseudo-second-order kinetic models were tested to fit the experimental data obtained for Dic and Caf uptake by AC. The kinetic study results are given in Table 4.

Table 4. Pseudo-first-order and pseudo-second-order parameters for adsorption of Dic and Caf onto AC based on argan nutshells.

	Pseudo First Order			Pseudo Second Order		
	Qe (mg/g)	K_1	R^2	Qe (mg/g)	K_2	R^2
Dic	41.01	−0.00016	0.991	91.16	77576.79	0.999
Caf	9.28	−0.00016	0.872	95.99	463867.18	0.999

The kinetic data of Dic and Caf adsorption on AC based on argan nutshells was investigated at temperatures of 30 °C. The best fitting model was defined by the higher determination coefficient (R^2). The pseudo-second-order model was the most suitable for the Dic and Caf adsorption on AC based on argan nutshells data because this model has a R^2 value close to 1 compared to pseudo-first-order model. The experimental adsorption capacity for Dic (91.16 mg/g) and for Caf (95.99 mg/g) was also close to the calculated adsorption capacity for Dic (82.60 mg/g) and for Caf (93.09 mg/g) (Figures 7 and 8). This suggests that the adsorption kinetics of emergent contaminants can be well described by the pseudo-second-order kinetic model. This means that the adsorption process is one of chemisorption with various interactions, such as electrostatic attractions, stacking (pi-stacking interactions (attractive, noncovalent interactions between aromatic rings)), hydrogen-bond formation, and Van der Waals forces between the adsorbent and adsorbate [88].

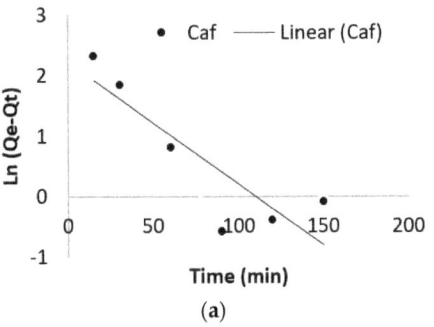
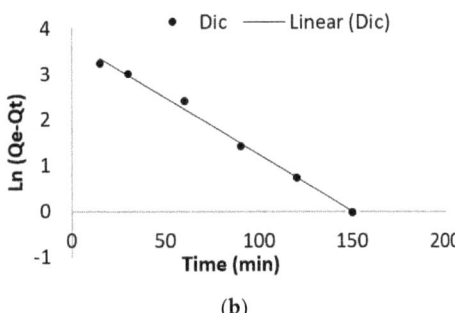

Figure 7. Pseudo-first-order kinetic model applied to the adsorption of Caf (**a**) and Dic (**b**) on activated carbon from argan nutshells.

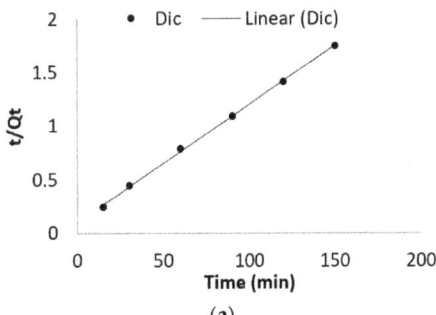
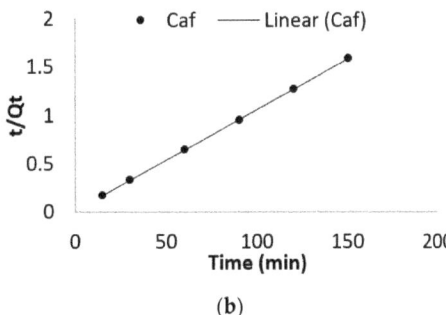

Figure 8. Pseudo-second-order kinetic model applied to the adsorption of Caf (**a**) and Dic (**b**) on activated carbon from argan nutshells.

3.2.2. Adsorption Isotherm

The adsorption isotherm study was done to describe the interactions between Dic and Caf on AC prepared from argan nutshells. It is important for the interpretation of the surface properties, the adsorption capacities of AC, and to complete the adsorption isotherm study that the equilibrium data were fitted to the Langmuir model and the Freundlich model [89,90]. The Langmuir and Freundlich parameters of Dic and Caf adsorption on AC were calculated using Equations (S5) and (S6) in the Supplementary information. The isotherm parameters are listed in Table 5, Based on the comparison of the correlation coefficient (R^2) values of Dic and Caf adsorbed on AC (Figures 9 and 10).

Table 5. Parameters of Langmuir and Freundlich models of Dic and Caf onto AC based on argan shells.

	Langmuir Isotherm				Freundlich Isotherm		
	Q_m(mg/g)	Kl (L/mg)	R^2	Rl	1/n	Kf (L/g)	R^2
Dic	126.16	0.24	0.99	0.17	1.50	38.19	0.85
Caf	210.65	0.05	0.99	0.27	1.08	61.43	0.97

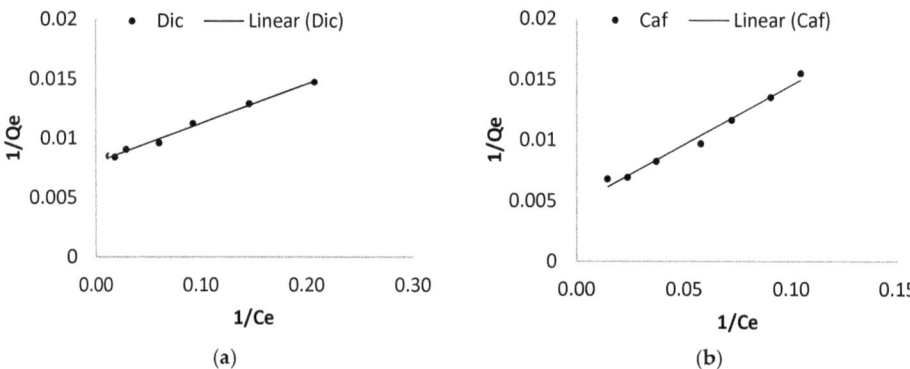

Figure 9. Langmuir isotherm of Dic (**a**) and Caf (**b**) on AC based on argan nutshells.

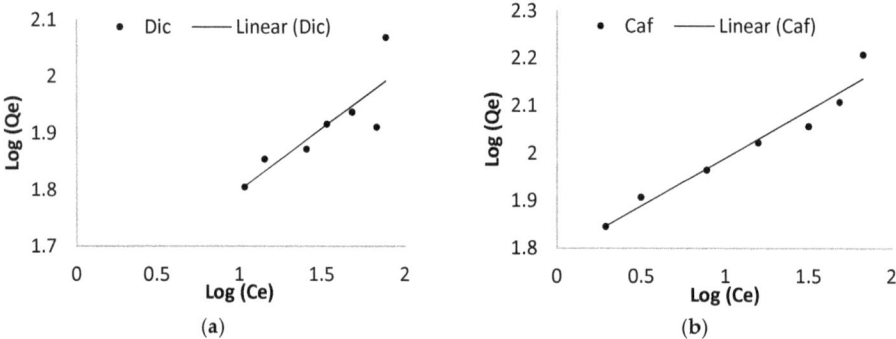

Figure 10. Freundlich isotherm of Dic (**a**) and Caf (**b**) on AC based on argan nutshells.

Figure 9 describes the linear equations 1/Qe versus 1/Ce of Dic and Caf on AC based on argan nutshells. The Qm and K*l* values are presented in Table 5, which show contaminants adsorption on the heat-resistant activated Langmuir angle and the calculated values of the parameters.

Based on Table 5, the correlation coefficient of the linear regression equation (R^2) of the Langmuir isotherm adsorption model is reasonable for the adsorption of Dic and Caf by activated carbon based on argan nutshells with values of 0.996, and 0.990, respectively. The maximum adsorption capacities (Qm) of Dic and Caf by activated carbon based on Argan nutshells calculated from the Langmuir model are 126.16 mg/g, and 210.65 mg/g, respectively.

When the experimental equilibrium data are appropriately described by the Langmuir model, it is essential to calculate the separation factor [91]. It was originally proposed that the essential characteristics of the Langmuir isotherm model could be indicated in terms of a dimensionless constant separation factor or equilibrium parameter R*l*, which is defined as follows:

$$Rl = \frac{1}{1 + (Kl*Co)} \quad (8)$$

where R*l* is a constant separation factor (dimensionless) of a solid–liquid adsorption system, K*l* is the Langmuir equilibrium constant, and Co is the initial concentration.

The results show that R*l* values for Dic were 0.169 and for Caf 0.266. All of the values between zero and one indicate the suitability of the Langmuir isotherm model for the description of the adsorption process of Dic and Caf.

Figure 10 describes the linear equation log (Qe) versus log (Ce), thereby determining the constants Kf and n, as shown in Table 5.

Table 5 shows the adsorption process of Dic and Caf on activated carbon based on argan nutshells, according to the Freundlich isotherm model with values of 0.85 and 0.97, respectively. These indicate that the Freundlich isotherm model is not suitable for describing the contaminants adsorption process by the adsorbents.

The results show that n values for Dic were 1.501 and for Caf were 1.076. They were both superior to one, indicating that the adsorption isotherms are poorly modelled by the Freundlich equation.

Furthermore, the Langmuir isotherm model has a higher regression coefficient R^2 than the Freundlich model (Table 5), indicating the Langmuir model provides a better description of AC (based on argan nutshells) adsorption process in Dic and Caf. Therefore, these results suggest monolayer adsorption of AC on the surface of the adsorbent.

Table 1 shows a comparison of absorbance capacity of Dic and Caf on various adsorbents reported in the literature, since the absorbance capacity of contaminants adsorbed varies as a function of different parameters (Initial concentration, contact time, etc.). Nevertheless, AC from argan nutshells presented high capacities for Dic and Caf, comparable or even higher than the ones obtained with other activated carbons derived from agricultural waste (Table 1).

To understand the mechanisms associated with the adsorption of Dic and Caf by the AC from argan nutshells, it is important to evaluate a potential practical application of adsorbents related to the removal of this type of contaminant.

The results presented in Table 6 also highlight that the surface area is not always the important feature in the removal of these adsorbate molecules.

Table 6. Adsorption capacities and surface area of different contaminants using AC based on argan fruits shells compared to the literature data.

Adsorbent	Contaminants	BET Surface Area (m^2/g)	Adsorption Capacity (mg/g)/Removal Efficiency (%)	References
AC based on Argan nutshells	Dic	1007	126	Present work
	Caf		210	
AC-HP	BPA	1372	1250	[92]
ACH	DCF	1542	149	[93]
	PARX		168	
ANS	BPA	42	1162	[94]
ANS	CV	-	98.21%	[95]

As mentioned before, textural properties were not the main factors in the adsorption of Dic and Caf since the AC obtained from argan nutshells presented a higher surface area and pore volume did not perform better regarding adsorption capacity of Dic and Caf. The large micropores developed on AC from argan nutshells do not provide an optimum size for adsorbates adsorption, which can explain the minor impact of surface area (Table 6). In fact, the role of the microporous network in the interaction with pharmaceutical molecules was previously demonstrated: If the critical dimension of the adsorbate molecule is close to the width of the micropores there will be an enhanced interaction and packing of the molecules [93].

3.2.3. Effect of Temperature and Thermodynamic Study

The effect of temperature on the adsorption phenomenon was studied by varying this parameter from 10 °C to 30 °C using a thermostat bath to maintain the temperature at the desired value. The tests were carried out by stirring 1 g of activated carbon based on argan shells with 100 mg of each contaminant (diclofenac and caffeine) in 1 L of the solution.

Initially, thermodynamic parameters such as Gibbs free energy ($\Delta G°$), enthalpy ($\Delta H°$), and entropy ($\Delta S°$) for diclofenac and caffeine adsorptions were determined by the slope and intercept in Ln(K) versus 1/T plot (Figure 11) that allowed for calculating the values of $\Delta H°$ and $\Delta S°$ in both matrices. The results are shown in Table 7.

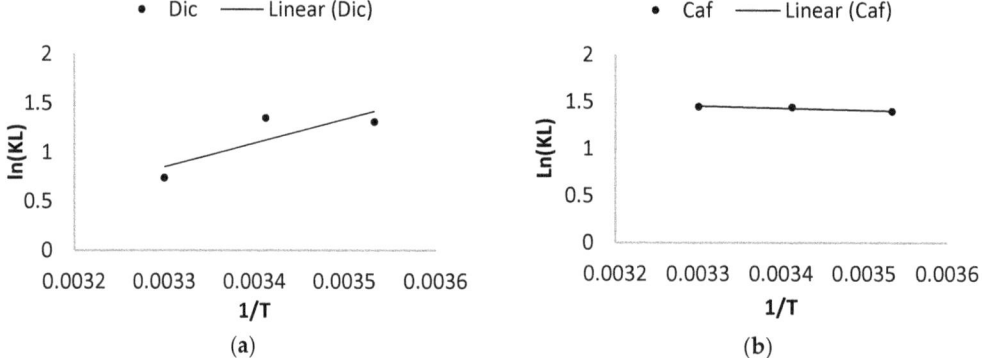

Figure 11. Plot of ln(Kc) versus temperature (1/T) for thermodynamic parameter calculation for the adsorption of Dic (**a**) and Caf (**b**) on AC based on argan nutshells.

Table 7. Thermodynamic parameters relating to the adsorption of contaminants (Dic and Caf) on activated carbon based on argan nutshells.

Contaminants	T (°C)	$\Delta G°$ (Kj/mol)	$\Delta H°$ (Kj/mol)	$\Delta S°$ (Kj/mol/k)	R^2
Dic	10	−3.28			
	20	−3.07	−20.11	−59.32	0.82
	30	−1.85			
Caf	10	−3.29			
	20	−3.51	1.77	17.95	0.92
	30	−3.64			

The negative values of the three parameters $\Delta H°$, $\Delta G°$, and $\Delta S°$ of diclofenac indicate that the reaction is spontaneous and exothermic and that the order of distribution of the contaminant molecules on the adsorbent is large compared to that in solution. Furthermore, an examination of the standard enthalpy values of the adsorption (<40 kJ/mol) shows that it is physisorption. In the case of diclofenac, the negative $\Delta S°$ value shows that adsorption occurs with increasing order at the solid–solution interface. The negative values of $\Delta G°$ increase with temperature and indicate an increase in disorder during adsorption, and the randomness increases at the solid–solution interface during this binding process. This can be explained by the redistribution of energy between the absorbent and the absorbate.

The positive value of $\Delta H°$ confirms the endothermic nature of the adsorption process of caffeine (values lower than 40 Kj/mol). Therefore, the adsorption regarding the matrices occurs by physisorption. Indeed, $\Delta S°$ presented positive values, which agrees with a dissociative mechanism. Moreover, the positive value of $\Delta S°$ shows the increased randomness at the solid–solution interface during the adsorption. It might display an increment of the degrees of freedom for the caffeine molecules in the solution. Additionally, Table 7 shows more negative $\Delta G°$ values as the temperature increased; these indicate that the adsorption process is spontaneous, and spontaneity increases with an increase in temperature.

3.3. Statistical Analysis

Results are reported as the means of four replicates. Data obtained were subjected to one-way analysis of variance (ANOVA) for assessing the significance of quantitative changes in the variables as a result of biochar treatments. The statistical analysis was done by the Statistical Package for Social Science (SPSS 23.0).

According to the statistical analysis (Table 8), the effect of the dose shows that there is a significant difference ($p > 0.05$) between the means of the adsorption capacities of caffeine and diclofenac by the activated carbons from argan nutshells (AC). On the other hand, the statistical analysis of the effect of the initial concentration shows that the test is significant at the 5% level. Furthermore, the statistical analysis of the effect of contact time shows that the test is highly significant at the 1% value; there is a significant difference between the mean adsorption capacities of caffeine and diclofenac by the activated carbons from argan nutshells (AC). Moreover, the statistical analysis of the effect of temperature shows that the highly significant test at the 5% threshold shows a significant difference between the means of adsorption capacities of caffeine and diclofenac by the activated carbons from argan nutshells (AC).

Table 8. Analysis of variance (F-test) of the effects on the adsorption of caffeine and diclofenac by the activated carbons from argan nutshells (AC).

Type of Analysis	Parameter Study	Type of Sample	Mean	Std. Error	95% Confidence Interval		Test ANOVA	
					Lower Bound	Upper Bound	F	Sig.
Effect of adsorbent dose on adsorption yield of Caffeine and Diclofenac	Adsorption yield, Caffeine (%)	AC	72.448	10.416	46.960	97.937	0.001	0.000 S
	Adsorption yield, Diclofenac (%)	AC	60.466	9.654	36.343	84.089	0.002	0.000 S
Effect of Concentration on the adsorption capacity of Caffeine and Diclofenac	Adsorption capacity, Caffeine (mg/g)	AC	79.509	15.820	11.438	147.579	0.002	0.000 S
	Adsorption capacity, Diclofenac (mg/g)	AC	80.226	12.080	28.247	132.206	0.001	0.000 S
Effect of contact time on adsorption capacity of Caffeine and Diclofenac	Adsorption capacity, Caffeine (mg/g)	AC	91.839	1.619	87.675	96.002	0.003	0.001 S
	Adsorption capacity, Diclofenac (mg/g)	AC	76.133	4.123	65.534	86.733	0.002	0.000 S
Effect of temperature on adsorption of Caffeine and Diclofenac	Adsorption capacity, Caffeine (mg/g)	AC	95.869	5.743	71.548	120.583	0.001	0.001 S
	Adsorption capacity, Caffeine (mg/g)	AC	83.449	4.569	63.786	103.112	0.003	0.000 S

Values are averages ± standard deviation of triplicate analysis. Data obtained were subjected to one-way analysis of variance (ANOVA). NS: Non-significant ($p > 0.05$). S: Significant ($p < 0.05$). AC: Activated carbons from argan nutshells.

4. Conclusions

The adsorption experiments show that the argan shells used were very effective in removing emerging contaminants such as diclofenac and caffeine at relatively low concentrations in aqueous medium. Adsorption tests showed that the equilibrium time was 60 and 90 min for Dic and Caf, respectively. The adsorption of Dic and Caf on activated carbon (AC) from argan nutshells is perfectly described by a pseudo-second-order kinetic model. The highest adsorption capacity determined by the mathematical model of Langmuir was about 126 mg/g for Dic and 210 mg/g for Caf. The thermodynamic parameters linked to the studied absorbent/adsorbate system show that the adsorption process is spontaneous and exothermic for diclofenac and endothermic for caffeine. Therefore, the chemical activation of argan shells improves its adsorption capacity. Thus, we can offer an adsorption material at low cost that can possibly contribute to the protection of the environment, especially in the purification of water. The valorization of Moroccan argan shells has been highlighted in this work.

Author Contributions: Conceptualization, B.B., D.C.-P. and F.E.M.; Funding acquisition, A.M.; Investigation, B.B., D.C.-P. and F.E.M.; Methodology, Y.E.Z., R.M., A.M., J.M.Q. and M.H.Z.; Supervision, Y.E.Z., A.M. and M.H.Z. All authors have read and agreed to the published version of the manuscript.

Funding: This research received no external funding.

Institutional Review Board Statement: Not applicable.

Informed Consent Statement: Not applicable.

Data Availability Statement: Not applicable.

Conflicts of Interest: The authors declare no conflict of interest.

References

1. Bo, L.; Shengen, Z.; Chang, C.-C. Emerging Pollutants—Part II: Treatment. *Water Environ. Res.* **2016**, *88*, 1876–1904. [CrossRef] [PubMed]
2. Tran, N.H.; Reinhard, M.; Khan, E.; Chen, H.; Nguyen, V.T.; Li, Y.; Goh, S.G.; Nguyen, Q.B.; Saeidi, N.; Gin, K.Y.-H. Emerging Contaminants in Wastewater, Stormwater Runoff, and Surface Water: Application as Chemical Markers for Diffuse Sources. *Sci. Total Environ.* **2019**, *676*, 252–267. [CrossRef] [PubMed]
3. Gomes, A.R.; Justino, C.; Rocha-Santos, T.; Freitas, A.C.; Duarte, A.C.; Pereira, R. Review of the Ecotoxicological Effects of Emerging Contaminants to Soil Biota. *J. Environ. Sci. Health A* **2017**, *52*, 992–1007. [CrossRef] [PubMed]
4. Álvarez, S.; Ribeiro, R.S.; Gomes, H.T.; Sotelo, J.L.; García, J. Synthesis of Carbon Xerogels and Their Application in Adsorption Studies of Caffeine and Diclofenac as Emerging Contaminants. *Chem. Eng. Res. Des.* **2015**, *95*, 229–238. [CrossRef]
5. Fraker, S.L.; Smith, G.R. Direct and Interactive Effects of Ecologically Relevant Concentrations of Organic Wastewater Contaminants on Rana Pipiens Tadpoles. *Environ. Toxicol.* **2004**, *19*, 250–256. [CrossRef]
6. Onaga Medina, F.M.; Aguiar, M.B.; Parolo, M.E.; Avena, M.J. Insights of Competitive Adsorption on Activated Carbon of Binary Caffeine and Diclofenac Solutions. *J. Environ. Manag.* **2021**, *278*, 111523. [CrossRef]
7. Santos-Silva, T.G.; Montagner, C.C.; Martinez, C.B.R. Evaluation of Caffeine Effects on Biochemical and Genotoxic Biomarkers in the Neotropical Freshwater Teleost *Prochilodus Lineatus*. *Environ. Toxicol. Pharmacol.* **2018**, *58*, 237–242. [CrossRef]
8. Oliveira, M.F.; da Silva, M.G.C.; Vieira, M.G.A. Equilibrium and Kinetic Studies of Caffeine Adsorption from Aqueous Solutions on Thermally Modified Verde-Lodo Bentonite. *Appl. Clay Sci.* **2019**, *168*, 366–373. [CrossRef]
9. Torrellas, S.Á.; García Lovera, R.; Escalona, N.; Sepúlveda, C.; Sotelo, J.L.; García, J. Chemical-Activated Carbons from Peach Stones for the Adsorption of Emerging Contaminants in Aqueous Solutions. *Chem. Eng. J.* **2015**, *279*, 788–798. [CrossRef]
10. Danish, M.; Birnbach, J.; Mohamad Ibrahim, M.N.; Hashim, R.; Majeed, S.; Tay, G.S.; Sapawe, N. Optimization Study of Caffeine Adsorption onto Large Surface Area Wood Activated Carbon through Central Composite Design Approach. *Environ. Nanotechnol. Monit. Manag.* **2021**, *16*, 100594. [CrossRef]
11. Danish, M. Application of Date Stone Activated Carbon for the Removal of Caffeine Molecules from Water. *Mater. Today Proc.* **2020**, *31*, 18–22. [CrossRef]
12. Zanella, H.G.; Spessato, L.; Lopes, G.K.P.; Yokoyama, J.T.C.; Silva, M.C.; Souza, P.S.C.; Ronix, A.; Cazetta, A.L.; Almeida, V.C. Caffeine Adsorption on Activated Biochar Derived from Macrophytes (*Eichornia Crassipes*). *J. Mol. Liq.* **2021**, *340*, 117206. [CrossRef]
13. da Silva Vasconcelos de Almeida, A.; Vieira, W.T.; Bispo, M.D.; de Melo, S.F.; da Silva, T.L.; Balliano, T.L.; Vieira, M.G.A.; Soletti, J.I. Caffeine Removal Using Activated Biochar from Açaí Seed (*Euterpe Oleracea Mart*): Experimental Study and Description of Adsorbate Properties Using Density Functional Theory (DFT). *J. Environ. Chem. Eng.* **2021**, *9*, 104891. [CrossRef]
14. Beltrame, K.K.; Cazetta, A.L.; de Souza, P.S.C.; Spessato, L.; Silva, T.L.; Almeida, V.C. Adsorption of Caffeine on Mesoporous Activated Carbon Fibers Prepared from Pineapple Plant Leaves. *Ecotoxicol. Environ. Saf.* **2018**, *147*, 64–71. [CrossRef]
15. Francoeur, M.; Ferino-Pérez, A.; Yacou, C.; Jean-Marius, C.; Emmanuel, E.; Chérémond, Y.; Jauregui-Haza, U.; Gaspard, S. Activated Carbon Synthetized from *Sargassum* (Sp) for Adsorption of Caffeine: Understanding the Adsorption Mechanism Using Molecular Modeling. *J. Environ. Chem. Eng.* **2021**, *9*, 104795. [CrossRef]
16. Galhetas, M.; Mestre, A.S.; Pinto, M.L.; Gulyurtlu, I.; Lopes, H.; Carvalho, A.P. Chars from Gasification of Coal and Pine Activated with K_2CO_3: Acetaminophen and Caffeine Adsorption from Aqueous Solutions. *J. Colloid Interface Sci.* **2014**, *433*, 94–103. [CrossRef]

17. Mengesha, D.N.; Abebe, M.W.; Appiah-Ntiamoah, R.; Kim, H. Ground Coffee Waste-Derived Carbon for Adsorptive Removal of Caffeine: Effect of Surface Chemistry and Porous Structure. *Sci. Total Environ.* **2021**, 151669. [CrossRef]
18. Melo, L.L.A.; Ide, A.H.; Duarte, J.L.S.; Zanta, C.L.P.S.; Oliveira, L.M.T.M.; Pimentel, W.R.O.; Meili, L. Caffeine Removal Using *Elaeis Guineensis* Activated Carbon: Adsorption and RSM Studies. *Environ. Sci. Pollut. Res.* **2020**, *27*, 27048–27060. [CrossRef]
19. Cunha, M.R.; Lima, E.C.; Cimirro, N.F.G.M.; Thue, P.S.; Dias, S.L.P.; Gelesky, M.A.; Dotto, G.L.; dos Reis, G.S.; Pavan, F.A. Conversion of *Eragrostis Plana* Nees Leaves to Activated Carbon by Microwave-Assisted Pyrolysis for the Removal of Organic Emerging Contaminants from Aqueous Solutions. *Environ. Sci. Pollut. Res.* **2018**, *25*, 23315–23327. [CrossRef]
20. Avcu, T.; Üner, O.; Geçgel, Ü. Adsorptive Removal of Diclofenac Sodium from Aqueous Solution onto Sycamore Ball Activated Carbon—Isotherms, Kinetics, and Thermodynamic Study. *Surf. Interfaces* **2021**, *24*, 101097. [CrossRef]
21. Naghipour, D.; Hoseinzadeh, L.; Taghavi, K.; Jaafari, J. Characterization, Kinetic, Thermodynamic and Isotherm Data for Diclofenac Removal from Aqueous Solution by Activated Carbon Derived from Pine Tree. *Data Brief* **2018**, *18*, 1082–1087. [CrossRef] [PubMed]
22. El Naga, A.O.; El Saied, M.; Shaban, S.A.; El Kady, F.Y. Fast Removal of Diclofenac Sodium from Aqueous Solution Using Sugar Cane Bagasse-Derived Activated Carbon. *J. Mol. Liq.* **2019**, *285*, 9–19. [CrossRef]
23. Saucier, C.; Adebayo, M.A.; Lima, E.C.; Cataluña, R.; Thue, P.S.; Prola, L.D.T.; Puchana-Rosero, M.J.; Machado, F.M.; Pavan, F.A.; Dotto, G.L. Microwave-assisted activated carbon from cocoa shell as adsorbent for removal of sodium diclofenac and nimesulide from aqueous effluents. *J. Hazard. Mater.* **2015**, *289*, 18–27. [CrossRef] [PubMed]
24. Malhotra, M.; Suresh, S.; Garg, A. Tea Waste Derived Activated Carbon for the Adsorption of Sodium Diclofenac from Wastewater: Adsorbent Characteristics, Adsorption Isotherms, Kinetics, and Thermodynamics. *Environ. Sci. Pollut. Res.* **2018**, *25*, 32210–32220. [CrossRef]
25. Bernardo, M.; Rodrigues, S.; Lapa, N.; Matos, I.; Lemos, F.; Batista, M.K.S.; Carvalho, A.P.; Fonseca, I. High Efficacy on Diclofenac Removal by Activated Carbon Produced from Potato Peel Waste. *Int. J. Environ. Sci. Technol.* **2016**, *13*, 1989–2000. [CrossRef]
26. Baccar, R.; Sarrà, M.; Bouzid, J.; Feki, M.; Blánquez, P. Removal of Pharmaceutical Compounds by Activated Carbon Prepared from Agricultural By-Product. *Chem. Eng. J.* **2012**, *211–212*, 310–317. [CrossRef]
27. Álvarez-Torrellas, S.; Muñoz, M.; Zazo, J.; Casas, J.A.; Garcia, M.M. Synthesis of High Surface Area Carbon Adsorbents Prepared from Pine Sawdust-*Onopordum Acanthium* L. for Nonsteroidal Anti-Inflammatory Drugs Adsorption. *J. Environ. Manag.* **2016**, *183*, 294–305. [CrossRef]
28. Vedenyapina, M.D.; Stopp, P.; Weichgrebe, D.; Vedenyapin, A.A. Adsorption of Diclofenac Sodium from Aqueous Solutions on Activated Carbon. *Solid Fuel Chem.* **2016**, *50*, 46–50. [CrossRef]
29. Fernandez, M.E.; Ledesma, B.; Román, S.; Bonelli, P.R.; Cukierman, A.L. Development and Characterization of Activated Hydrochars from Orange Peels as Potential Adsorbents for Emerging Organic Contaminants. *Bioresour. Technol.* **2015**, *183*, 221–228. [CrossRef]
30. Ferreira, L.; Rosales, E.; Danko, A.S.; Sanromán, M.A.; Pazos, M.M. *Bacillus Thuringiensis* a Promising Bacterium for Degrading Emerging Pollutants. *Process Saf. Environ. Prot.* **2016**, *101*, 19–26. [CrossRef]
31. Barrios, J.A.; Cano, A.; Becerril, J.E.; Jiménez, B. Influence of Solids on the Removal of Emerging Pollutants in Electrooxidation of Municipal Sludge with Boron-Doped Diamond Electrodes. *J. Electroanal. Chem.* **2016**, *776*, 148–151. [CrossRef]
32. Acero, J.L.; Benitez, F.J.; Real, F.J.; Rodriguez, E. Elimination of Selected Emerging Contaminants by the Combination of Membrane Filtration and Chemical Oxidation Processes. *Water Air Soil Pollut.* **2015**, *226*, 139. [CrossRef]
33. Esmaeeli, F.; Gorbanian, S.A.; Moazezi, N. Removal of Estradiol Valerate and Progesterone Using Powdered and Granular Activated Carbon from Aqueous Solutions. *Int. J. Environ. Res.* **2017**, *11*, 695–705. [CrossRef]
34. Leite, A.B.; Saucier, C.; Lima, E.C.; dos Reis, G.S.; Umpierres, C.S.; Mello, B.L.; Shirmardi, M.; Dias, S.L.P.; Sampaio, C.H. Activated Carbons from Avocado Seed: Optimisation and Application for Removal of Several Emerging Organic Compounds. *Environ. Sci. Pollut. Res.* **2018**, *25*, 7647–7661. [CrossRef] [PubMed]
35. Wong, S.; Lim, Y.; Ngadi, N.; Mat, R.; Hassan, O.; Inuwa, I.M.; Mohamed, N.B.; Low, J.H. Removal of Acetaminophen by Activated Carbon Synthesized from Spent Tea Leaves: Equilibrium, Kinetics and Thermodynamics Studies. *Powder Technol.* **2018**, *338*, 878–886. [CrossRef]
36. Gupta, V.K. Suhas Application of Low-Cost Adsorbents for Dye Removal—A Review. *J. Environ. Manag.* **2009**, *90*, 2313–2342. [CrossRef]
37. Ahmad, A.; Rafatullah, M.; Sulaiman, O.; Ibrahim, M.H.; Chii, Y.Y.; Siddique, B.M. Removal of Cu(II) and Pb(II) Ions from Aqueous Solutions by Adsorption on Sawdust of Meranti Wood. *Desalination* **2009**, *247*, 636–646. [CrossRef]
38. Ahmad, A.; Rafatullah, M.; Danish, M. Sorption Studies of Zn(II)- and Cd(II)Ions from Aqueous Solution on Treated Sawdust of Sissoo Wood. *Holz Als Roh-Und Werkst.* **2007**, *65*, 429–436. [CrossRef]
39. Dąbrowski, A. Adsorption—From Theory to Practice. *Adv. Colloid Interface Sci.* **2001**, *93*, 135–224. [CrossRef]
40. Rafatullah, M.; Sulaiman, O.; Hashim, R.; Ahmad, A. Adsorption of Copper (II), Chromium (III), Nickel (II) and Lead (II) Ions from Aqueous Solutions by Meranti Sawdust. *J. Hazard. Mater.* **2009**, *170*, 969–977. [CrossRef]
41. Rafatullah, M.; Sulaiman, O.; Hashim, R.; Ahmad, A. Adsorption of Methylene Blue on Low-Cost Adsorbents: A Review. *J. Hazard. Mater.* **2010**, *177*, 70–80. [CrossRef] [PubMed]
42. Guillot, J.-M.; Fernandez, B.; Le Cloirec, P. Advantages and Limits of Adsorption Sampling for Physico-Chemical Measurements of Odorous Compounds. *Analusis* **2000**, *28*, 180–187. [CrossRef]

43. Kumar, P.S.; Joshiba, G.J.; Femina, C.C.; Varshini, P.; Priyadharshini, S.; Karthick, M.A.; Jothirani, R. A Critical Review on Recent Developments in the Low-Cost Adsorption of Dyes from Wastewater. *Desalin. Water Treat.* **2019**, *172*, 395–416. [CrossRef]
44. Prasannamedha, G.; Kumar, P.S.; Mehala, R.; Sharumitha, T.J.; Surendhar, D. Enhanced Adsorptive Removal of Sulfamethoxazole from Water Using Biochar Derived from Hydrothermal Carbonization of Sugarcane Bagasse. *J. Hazard. Mater.* **2021**, *407*, 124825. [CrossRef] [PubMed]
45. Li, D.; Chen, L.; Zhang, X.; Ye, N.; Xing, F. Pyrolytic Characteristics and Kinetic Studies of Three Kinds of Red Algae. *Biomass Bioenergy* **2011**, *35*, 1765–1772. [CrossRef]
46. Chen, Y.; Zhu, Y.; Wang, Z.; Li, Y.; Wang, L.; Ding, L.; Gao, X.; Ma, Y.; Guo, Y. Application Studies of Activated Carbon Derived from Rice Husks Produced by Chemical-Thermal Process—A Review. *Adv. Colloid Interface Sci.* **2011**, *163*, 39–52. [CrossRef]
47. Ioannidou, O.; Zabaniotou, A. Agricultural Residues as Precursors for Activated Carbon Production—A Review. *Renew. Sustain. Energy Rev.* **2007**, *11*, 1966–2005. [CrossRef]
48. Kumar, B.G.; Shivakamy, K.; Miranda, L.R.; Velan, M. Preparation of Steam Activated Carbon from Rubberwood Sawdust (*Hevea Brasiliensis*) and Its Adsorption Kinetics. *J. Hazard. Mater.* **2006**, *136*, 922–929. [CrossRef]
49. Molina-Sabio, M.; Rodríguez-Reinoso, F. Role of Chemical Activation in the Development of Carbon Porosity. *Colloids Surf. A Physicochem. Eng. Asp.* **2004**, *241*, 15–25. [CrossRef]
50. Rivera-Utrilla, J.; Sánchez-Polo, M.; Gómez-Serrano, V.; Álvarez, P.M.; Alvim-Ferraz, M.C.M.; Dias, J.M. Activated Carbon Modifications to Enhance Its Water Treatment Applications. An Overview. *J. Hazard. Mater.* **2011**, *187*, 1–23. [CrossRef]
51. Lee, J.; Kim, J.; Hyeon, T. Recent Progress in the Synthesis of Porous Carbon Materials. *Adv. Mater.* **2006**, *18*, 2073–2094. [CrossRef]
52. Ahmedna, M.; Marshall, W.E.; Husseiny, A.A.; Rao, R.M.; Goktepe, I. The Use of Nutshell Carbons in Drinking Water Filters for Removal of Trace Metals. *Water Res.* **2004**, *38*, 1062–1068. [CrossRef] [PubMed]
53. Lanzetta, M.; Di Blasi, C. Pyrolysis Kinetics of Wheat and Corn Straw. *J. Anal. Appl. Pyrolysis* **1998**, *44*, 181–192. [CrossRef]
54. Minkova, V.; Razvigorova, M.; Bjornbom, E.; Zanzi, R.; Budinova, T.; Petrov, N. Effect of Water Vapour and Biomass Nature on the Yield and Quality of the Pyrolysis Products from Biomass. *Fuel Processing Technol.* **2001**, *70*, 53–61. [CrossRef]
55. Haykiri-Acma, H.; Yaman, S.; Kucukbayrak, S. Gasification of Biomass Chars in Steam–Nitrogen Mixture. *Energy Convers. Manag.* **2006**, *47*, 1004–1013. [CrossRef]
56. Cetin, E.; Moghtaderi, B.; Gupta, R.; Wall, T.F. Influence of Pyrolysis Conditions on the Structure and Gasification Reactivity of Biomass Chars. *Fuel* **2004**, *83*, 2139–2150. [CrossRef]
57. Aygün, A.; Yenisoy-Karakaş, S.; Duman, I. Production of Granular Activated Carbon from Fruit Stones and Nutshells and Evaluation of Their Physical, Chemical and Adsorption Properties. *Microporous Mesoporous Mater.* **2003**, *66*, 189–195. [CrossRef]
58. Tsai, W.T.; Chang, C.Y.; Lee, S.L. Preparation and Characterization of Activated Carbons from Corn Cob. *Carbon* **1997**, *35*, 1198–1200. [CrossRef]
59. Savova, D.; Apak, E.; Ekinci, E.; Yardim, F.; Petrov, N.; Budinova, T.; Razvigorova, M.; Minkova, V. Biomass Conversion to Carbon Adsorbents and Gas. *Biomass Bioenergy* **2001**, *21*, 133–142. [CrossRef]
60. Girgis, B.S.; Yunis, S.S.; Soliman, A.M. Characteristics of Activated Carbon from Peanut Hulls in Relation to Conditions of Preparation. *Mater. Lett.* **2002**, *57*, 164–172. [CrossRef]
61. Lua, A.C.; Yang, T.; Guo, J. Effects of Pyrolysis Conditions on the Properties of Activated Carbons Prepared from Pistachio-Nut Shells. *J. Anal. Appl. Pyrolysis* **2004**, *72*, 279–287. [CrossRef]
62. Ahmedna, M.; Marshall, W.E.; Rao, R.M. Production of Granular Activated Carbons from Select Agricultural By-Products and Evaluation of Their Physical, Chemical and Adsorption. *Bioresour. Technol.* **2000**, *71*, 113–123. [CrossRef]
63. Mansouri, F.E.; Farissi, H.E.; Zerrouk, M.H.; Cacciola, F.; Bakkali, C.; Brigui, J.; Lovillo, M.P.; da Silva, J.C.G. Dye Removal from Colored Textile Wastewater Using Seeds and Biochar of Barley (*Hordeum vulgare* L.). *Appl. Sci.* **2021**, *11*, 5125. [CrossRef]
64. Zhang, T.; Walawender, W.P.; Fan, L.T.; Fan, M.; Daugaard, D.; Brown, R.C. Preparation of Activated Carbon from Forest and Agricultural Residues through CO_2 Activation. *Chem. Eng. J.* **2004**, *105*, 53–59. [CrossRef]
65. Khallouki, F.; Haubner, R.; Ricarte, I.; Erben, G.; Klika, K.; Ulrich, C.M.; Owen, R.W. Identification of Polyphenolic Compounds in the Flesh of Argan (Morocco) Fruits. *Food Chem.* **2015**, *179*, 191–198. [CrossRef] [PubMed]
66. Matthäus, B.; Guillaume, D.; Gharby, S.; Haddad, A.; Harhar, H.; Charrouf, Z. Effect of Processing on the Quality of Edible Argan Oil. *Food Chem.* **2010**, *120*, 426–432. [CrossRef]
67. Dahbi, M.; Kiso, M.; Kubota, K.; Horiba, T.; Chafik, T.; Hida, K.; Matsuyama, T.; Komaba, S. Synthesis of Hard Carbon from Argan Shells for Na-Ion Batteries. *J. Mater. Chem. A* **2017**, *5*, 9917–9928. [CrossRef]
68. Chafik, T. Matériaux Carbonés Nanoporeux Préparés à Partir de La Coque Du Fruit d'argan. International Patent Application No. PCT/MA2011/000009. WO Patent 2012050411A1, 19 April 2012.
69. Edebali, S. *Advanced Sorption Process Applications*; IntechOpen: London, UK, 2019; ISBN 978-1-78984-819-9.
70. Kumar, P.; Ramalingam, S.; Senthamarai, C.; Niranjanaa, M.; Vijayalakshmi, P.; Sivanesan, S. Adsorption of Dye from Aqueous Solution by Cashew Nut Shell: Studies on Equilibrium Isotherm, Kinetics and Thermodynamics of Interactions. *Desalination* **2010**, *261*, 52–60. [CrossRef]
71. Lagergren, S.K. About the Theory of So-Called Adsorption of Soluble Substances. *Sven. Vetenskapsakad. Handingarl* **1898**, *24*, 1–39.
72. Yuh-Shan, H. Citation Review of Lagergren Kinetic Rate Equation on Adsorption Reactions. *Scientometrics* **2004**, *59*, 171–177. [CrossRef]
73. Ho, Y. The Kinetics of Sorption of Divalent Metal Ions onto Sphagnum Moss Peat. *Water Res.* **2000**, *34*, 735–742. [CrossRef]

74. Limousin, G.; Gaudet, J.-P.; Charlet, L.; Szenknect, S.; Barthès, V.; Krimissa, M. Sorption Isotherms: A Review on Physical Bases, Modeling and Measurement. *Appl. Geochem.* **2007**, *22*, 249–275. [CrossRef]
75. Foo, K.Y.; Hameed, B.H. An Overview of Dye Removal via Activated Carbon Adsorption Process. *Desalination Water Treat.* **2010**, *19*, 255–274. [CrossRef]
76. Al-Ghouti, M.A.; Da'ana, D.A. Guidelines for the Use and Interpretation of Adsorption Isotherm Models: A Review. *J. Hazard. Mater.* **2020**, *393*, 122383. [CrossRef] [PubMed]
77. Allen, S.J.; Mckay, G.; Porter, J.F. Adsorption Isotherm Models for Basic Dye Adsorption by Peat in Single and Binary Component Systems. *J. Colloid Interface Sci.* **2004**, *280*, 322–333. [CrossRef] [PubMed]
78. Ayawei, N.; Ebelegi, A.N.; Wankasi, D. Modelling and Interpretation of Adsorption Isotherms. *J. Chem.* **2017**, *2017*, e3039817. [CrossRef]
79. Kecili, R.; Hussain, C.M. Chapter 4—Mechanism of Adsorption on Nanomaterials. In *Nanomaterials in Chromatography*; Hussain, C.M., Ed.; Elsevier: Amsterdam, The Netherlands, 2018; pp. 89–115. ISBN 978-0-12-812792-6.
80. Bulut, Y.; Aydın, H. A Kinetics and Thermodynamics Study of Methylene Blue Adsorption on Wheat Shells. *Desalination* **2006**, *194*, 259–267. [CrossRef]
81. Gök, Ö.; Özcan, A.; Erdem, B.; Özcan, A.S. Prediction of the Kinetics, Equilibrium and Thermodynamic Parameters of Adsorption of Copper(II) Ions onto 8-Hydroxy Quinoline Immobilized Bentonite. *Colloids Surf. A Physicochem. Eng. Asp.* **2008**, *317*, 174–185. [CrossRef]
82. Tran, H.N.; You, S.-J.; Chao, H.-P. Thermodynamic Parameters of Cadmium Adsorption onto Orange Peel Calculated from Various Methods: A Comparison Study. *J. Environ. Chem. Eng.* **2016**, *4*, 2671–2682. [CrossRef]
83. Ramesh, A.; Lee, D.J.; Wong, J.W.C. Thermodynamic Parameters for Adsorption Equilibrium of Heavy Metals and Dyes from Wastewater with Low-Cost Adsorbents. *J. Colloid Interface Sci.* **2005**, *291*, 588–592. [CrossRef]
84. Liu, X.; Lee, D.-J. Thermodynamic Parameters for Adsorption Equilibrium of Heavy Metals and Dyes from Wastewaters. *Bioresour. Technol.* **2014**, *160*, 24–31. [CrossRef] [PubMed]
85. Anastopoulos, I.; Kyzas, G.Z. Are the Thermodynamic Parameters Correctly Estimated in Liquid-Phase Adsorption Phenomena? *J. Mol. Liq.* **2016**, *218*, 174–185. [CrossRef]
86. Burhenne, L.; Messmer, J.; Aicher, T.; Laborie, M.-P. The Effect of the Biomass Components Lignin, Cellulose and Hemicellulose on TGA and Fixed Bed Pyrolysis. *J. Anal. Appl. Pyrolysis* **2013**, *101*, 177–184. [CrossRef]
87. Thommes, M.; Kaneko, K.; Neimark, A.V.; Olivier, J.P.; Rodriguez-Reinoso, F.; Rouquerol, J.; Sing, K.S.W. Physisorption of Gases, with Special Reference to the Evaluation of Surface Area and Pore Size Distribution (IUPAC Technical Report). *Pure Appl. Chem.* **2015**, *87*, 1051–1069. [CrossRef]
88. Tran, T.H.; Le, A.H.; Pham, T.H.; Nguyen, D.T.; Chang, S.W.; Chung, W.J.; Nguyen, D.D. Adsorption Isotherms and Kinetic Modeling of Methylene Blue Dye onto a Carbonaceous Hydrochar Adsorbent Derived from Coffee Husk Waste. *Sci. Total Environ.* **2020**, *725*, 138325. [CrossRef]
89. Langmuir, I. The adsorption of gases on plane surfaces of glass, mica and platinum. *J. Am. Chem. Soc.* **1918**, *40*, 1361–1403. [CrossRef]
90. Freundlich, H. Über die Adsorption in Lösungen. *Z. Phys. Chem.* **1907**, *57U*, 385–470. [CrossRef]
91. Hall, K.R.; Eagleton, L.C.; Acrivos, A.; Vermeulen, T. Pore- and Solid-Diffusion Kinetics in Fixed-Bed Adsorption under Constant-Pattern Conditions. *Ind. Eng. Chem. Fund.* **1966**, *5*, 212–223. [CrossRef]
92. Zbair, M.; Ainassaari, K.; Drif, A.; Ojala, S.; Bottlinger, M.; Pirilä, M.; Bensitel, M.; Brahmi, R.; Keiski, R.L. Toward new benchmark adsorbents: Preparation and characterization of activated carbon from argan nut shell for bisphenol A removal. *Environ. Sci. Pollut. Res.* **2018**, *25*, 1869–1882. [CrossRef]
93. Mokhati, A.; Benturki, O.; Bernardo, M.; Kecira, Z.; Matos, I.; Lapa, N.; Ventura, M.; Soares, O.; Rego, A.B.D.; Fonseca, I. Nanoporous carbons prepared from argan nutshells as potential removal agents of diclofenac and paroxetine. *J. Mol. Liq.* **2021**, *326*, 115368. [CrossRef]
94. Zbair, M.; Bottlinger, M.; Ainassaari, K.; Ojala, S.; Stein, O.; Keiski, R.L.; Bensitel, M.; Brahmi, R. Hydrothermal Carbonization of Argan Nut Shell: Functional Mesoporous Carbon with Excellent Performance in the Adsorption of Bisphenol A and Diuron. *Waste Biomass Valor.* **2020**, *11*, 1565–1584. [CrossRef]
95. El Khomri, M.; El Messaoudi, N.; Dbik, A.; Bentahar, S.; Lacherai, A.; Faska, N.; Jada, A. Regeneration of Argan Nutshell and Almond Shell Using HNO_3 for Their Reusability to Remove Cationic Dye from Aqueous Solution. *Chem. Eng. Commun.* **2021**, 1963960. [CrossRef]

Article

Preparation and Modification of Activated Carbon for the Removal of Pharmaceutical Compounds via Adsorption and Photodegradation Processes: A Comparative Study

Brahim Samir [1], Nabil Bouazizi [1], Patrick Nkuigue Fotsing [1], Julie Cosme [1], Veronique Marquis [1], Guilherme Luiz Dotto [1,2], Franck Le Derf [1] and Julien Vieillard [1,*]

[1] Normandie Université, UNIROUEN, INSA Rouen, CNRS, COBRA (UMR 6014), 27000 Evreux, France; brahim.samir@univ-rouen.fr (B.S.); bouazizi.nabil@hotmail.fr (N.B.); fotsing_p@yahoo.fr (P.N.F.); julie.cosme@univ-rouen.fr (J.C.); veronique.marquis@univ-rouen.fr (V.M.); guilherme_dotto@yahoo.com.br (G.L.D.); franck.lederf@univ-rouen.fr (F.L.D.)

[2] Research Group on Adsorptive and Catalytic Process Engineering (ENGEPAC), Federal University of Santa Maria, Av. Roraima, 1000-7, Santa Maria 97105-900, RS, Brazil

* Correspondence: julien.vieillard@univ-rouen.fr; Tel.: +33-232-291-598

Abstract: In the present research, the removal of pharmaceutical contaminants based on atenolol (AT) and propranolol (PR) using modified activated carbon (AC) in a liquid solution was studied. Two methods, adsorption and photodegradation, were used to eliminate AT and PR. First, AC was prepared from date stems and then modified via hydroxylation (AC-OH) and impregnated into titanium dioxide (AC-TiO$_2$) separately. The removal of AT and PR was investigated in terms of experimental parameters, such as pH, concentration, temperature, and the effectiveness of the processes. The results show that the removal of AT and PR reached 92% for the adsorption method, while 94% was registered for the photodegradation process. Likewise, in optimal experimental conditions, the adsorption of AT and PR over AC-OH showed good stability and recyclability, achieving five cycles without a visible decrease in the removal capacity. The results obtained in this work suggest that the low-cost and environmentally friendly synthesis of AC-OH is suitable to be considered for wastewater treatment at the industrial scale. Interestingly, the above results open a potential pathway to determine whether adsorption or photodegradation is more suitable for eliminating wastewater-related pharmaceutical pollutants. Accordingly, the experimental results recommend adsorption as a promising, durable, eco-friendly wastewater treatment method.

Keywords: hydroxylation; titanium dioxide; adsorption; photodegradation; atenolol; propranolol

Citation: Samir, B.; Bouazizi, N.; Nkuigue Fotsing, P.; Cosme, J.; Marquis, V.; Dotto, G.L.; Le Derf, F.; Vieillard, J. Preparation and Modification of Activated Carbon for the Removal of Pharmaceutical Compounds via Adsorption and Photodegradation Processes: A Comparative Study. *Appl. Sci.* **2023**, *13*, 8074. https://doi.org/10.3390/app13148074

Academic Editors: Amanda Laca Pérez and Yolanda Patiño

Received: 20 May 2023
Revised: 23 June 2023
Accepted: 26 June 2023
Published: 11 July 2023

Copyright: © 2023 by the authors. Licensee MDPI, Basel, Switzerland. This article is an open access article distributed under the terms and conditions of the Creative Commons Attribution (CC BY) license (https://creativecommons.org/licenses/by/4.0/).

1. Introduction

In recent years, the concentrations of toxic pharmaceutical products in the global environment have increased substantially [1,2]. Pharmaceutical and chemical personal care products are widely used in daily life. Unfortunately, more than 50% of these hazardous products are discharged into the environment, such as in rivers, which can cause danger to fauna and flora [3]. Therefore, their destruction of ecological processes and functions in freshwater ecosystems is often called to be limited or reduced, as the continuous input of these pharmaceutical molecules in the water environment affects water safety, resulting in chronic toxic effects on organisms [4–7] and has potential impacts on human health through the food chain [8]. Numerous pharmaceutical products are discharged into the water environment, and atenolol (AT) and propranolol (PR) are the most used medicaments. PR and AT are medications called beta-blockers and are used to treat cardiovascular diseases, such as hypertension, tachycardia, and acute myocardial infarction [9]. In recent research studies, AT and PR were detected at high concentration levels in urban wastewater treatment plant effluents [10–13]. In addition, AT and PR were widely detected in hospital sewage and wastewater treatments in concentrations ranging from about 0.78 to 6.6 µg/L in Greece [14]. Regarding the above problems, water quality is in danger, and alternative reserves or crucial solutions are requested to alleviate this issue. Wastewater treatments via biological, filtration, settling, adsorption, coagulation, and many other processes displayed interesting results in removing these toxic products. However, almost all these processes showed a real weakness due to the lack of continuous properties and the high energy consumption. Adsorption is a very useful method for wastewater treatments, representing an eco-friendly and low-cost option. While adsorption is suitable for water treatment, it can only be considered an effective process if the adsorbent has complementary properties that ensure its eco-friendly and high adsorption capacities during its utilization.

Numerous adsorbents have been developed with the above aims, such as metallic oxide, biomass, activated carbon, graphene oxide, metal–organic frameworks, and zeolites [15–19]. To produce eco-friendly and low-cost properties, researchers focused on biomass and its derivatives as potential adsorbents and effective candidates for wastewater treatments. Up to now, activated carbon has been considered one of the best materials for water treatment due to its surface properties, low cost, and high capacity to remove pollutants from water. However, the rapid saturation of these adsorbents means they must be changed or recycled frequently; heterogeneous photocatalysts can be a good alternative to adsorption. This chemical process of photocatalysis involves reactive radical species, such as hydroxyl radicals (HO·), in the presence of a semiconductor catalyst based on a metal oxide to degrade the pharmaceutical molecules. Titanium dioxide (TiO_2) is a good catalyst due to its photochemical stability. Lu et al. and many other researchers investigated the immobilization of TiO_2 nanoparticles on an activated carbon surface to improve the photocatalytic activity and make the separation of treated effluent more effective [20–23]. The results obtained toward this goal are of great importance and, up to now, have not been developed. The coating of surfaces with TiO_2 produces a relatively low improvement regarding the photocatalytic reaction because of the particles' low dispersion and limited mass transfer between the pollutant molecules and the catalyst [24,25]. Catalysts can be more effective and easily separated from the effluent [26–28].

In this regard, activated carbon covered with TiO_2 semiconductors is active since it enhances the photocatalytic reaction between TiO_2 and the contaminants due to the adsorption of pollutants on its surface [29,30]. Increased adsorption contributes to a higher concentration of contaminants around the TiO_2 active sites [31]. Therefore, this study was designed to examine removing AT and GT pharmaceutical products from the water via adsorption and photodegradation with activated carbon and activated carbon covered with TiO_2. This first study focused on treating products with environmentally activated carbon and TiO_2. The results of this study could be used as a solution for water treatments, which is considered the most important environmental pollution issue to be resolved. In detail, this study evaluated the removal of AT and PR via an adsorption process using agro-waste (date stems) as a source of activated carbon (AC). At first, the adsorption of AT and PR was evaluated on AC and AC-OH, and then, their photodegradation in the presence of a heterogenous photocatalyst (AC-TiO_2) was tested.

2. Materials and Methods

2.1. Materials

Zinc chloride ($ZnCl_2$), hydrochloric acid (HCl) 37%, titanium (IV) isopropoxide (TTIP) 97%, atenolol (AT), propranolol (PR), acid nitric (HNO_3), hydrogen peroxide (H_2O_2), iron (III) nitrate nanohydrate ($Fe(NO_3)_3 \cdot 9H_2O$), and commercial activated carbon (AC) were purchased from Sigma-Aldrich, St. Louis, MI, USA. The chemical formulas and descriptions of AT and PR are presented in Table S1 and Figure S1.

2.2. Materials Synthesis

According to our previously published work, activated carbon was obtained from agro-waste, specifically from date stems [32,33]. The date stems were washed with distilled water, dried at 105 °C for 24 h, and cut at around 0.2 cm. For activation, 48 g of date stems were stirred with 96 g of ZnCl$_2$ (m(ZnCl$_2$) = 2 × m(AC)) for 4 h and carbonized at 600 °C for 2 h under dried air. Afterward, the obtained product was washed with HCl 37% and dried at 80 °C overnight. The product obtained wais dentified as AC. To increase the number of the hydroxyl group (-OH) at the surface of activated carbon (AC), AC was impregnated into a solution of hydrogen peroxide, nitric acid, and deionized water (1:1:5 $v/v/v$, respectively) at 70 °C for 4 h. The obtained powder was filtrated, washed with deionized water, and dried at 80 °C overnight. The resulting material is denoted as AC-OH a refers to OH-enriched activated carbon. In parallel, activated carbon was coated with a TiO$_2$ solution via an in situ impregnation method, according to a method described elsewhere [34]. Briefly, 0.1 g of AC is added to different quantities of TTIP (30, 50, and 70%) and dissolved in 20 mL of isopropanol. The mixture was stirred for 1 h, washed with isopropanol, and dried at 70 °C for 6 h. The final obtained powder is identified as activated carbon over titanium dioxide (AC-TiO$_2$).

2.3. Materials Characterization

The prepared products were characterized by scanning electron microscopy (SEM) using a JCM-6000 electron microscope (JEOL, Rueil Malmaison, France) to observe the surface morphology of AC and AC-OH. The chemical composition of AC before and after adsorption and AC-OH were characterized through Fourier transform IR spectroscopy (FTIR) using a Tensor 27 (Bruker, Wissembourg, France) spectrometer with a ZnSe ATR crystal device. For each spectrum, 20 scans were accumulated with a resolution of 4 cm^{-1}. The thermal stability of AC and AC-OH was analyzed by differential scanning calorimetry (DSC) using a DSC-92 (Setaram, Caluire et Cuire, France) device at a heating rate of 5 °C min^{-1} from room temperature to 550 °C. The zeta potential of each sample dispersion was measured in phase analysis light scattering (PALS) mode using a Zeta sizernanoZS setup (Malvern, Palaiseau). For the zeta measurements, nanoparticle suspension was obtained by adding 100 mg of each sample to 10 mL of ultrapure water.

2.4. Adsorption Experiments

The batch adsorption experiments were carried out using 50 mL of the solution containing a 50 mg L^{-1} concentration of the pharmaceutical molecules. The effect of pH was studied in the range from 2 to 10, where HCl and NaOH solution adjusts pH. Afterward, a predetermined amount of the adsorbent (0.010–0.8 g) was mixed into the solution before sonication at ambient temperature for 5–180 min. The supernatant was centrifuged at 2000 rpm for 3 min. Then, the absorbance was measured using a UV–vis spectrophotometer (Shimadzu Uv-1900) at λ_{max} of each molecule at 224 and 288 nm for AT and PR, respectively. The results are averages of a minimum of 3 experiments.

The percent removal of contaminant and the adsorption capacity was calculated using Equations (1) and (2).

$$\text{Contaminant removal \%} = \frac{C_0 - C_t}{C_0} \times 100 \quad (1)$$

$$\text{The capacity for adsorption} = (C_0 - C_t) \times \frac{V}{M} \quad (2)$$

C_0 mg/L and C_t mg/L are the initial concentration and concentration at time "t", respectively. "V" (mL) is the volume of (PR, AT), and "m" (mg) is the mass of the adsorbent.

2.5. Photocatalytic Experiments

The UV chamber (Model, 2000), 12.7 cm × 12.7 cm, was purchased from DYMAX, Germany. A UV lamp (DYMAX, Wiesbaden, Germany) that had a 400 W UV–mercury lamp. It generated a continuous spectrum of 320–400 nm with a measuring intensity of 225 mW/cm^2. Batch experiments were carried out in 500 mL quartz beakers, where the light source was held perpendicular to the batch reactor. Both catalytic and noncatalytic degradation was measured under identical conditions. Before absorption analysis, samples were collected at regular intervals and centrifuged (5000 rpm, 10 min). Then, the concentration of AT and PR were determined by measuring the absorbance (spectrophotometer UV-1900 Shimadzu, Marne La Vallée, France) at 224 and 288 nm, respectively.

3. Results

3.1. Morphological Properties

Figure 1 shows SEM analysis of AC (activated carbon), AC-OH (OH-enriched activated carbon), and AC-TiO$_2$ (activated carbon over titanium dioxide) samples. A smooth and nonuniform surface is registered for AC and its modified counterpart. The average particle size is around 50 µm. The presence of many cavities on the material's surface designs the morphology of AC. However, the addition of TiO$_2$ nanoparticles is confirmed by the high distribution of TiO$_2$ particles on the AC surface. A visible change is observed by the cavity displayed in Figure 1g, explaining the successful coating of TiO$_2$ particles onto the AC surface. The TiO$_2$ particles are spherical and distributed uniformly at the AC surface. It should be noted that the in situ synthesis of metallic particles presents a weak aggregation. Herein, immobilizing a high amount of TiO$_2$ over AC could be baneful for surface reactivity and the adsorption of pollutants. The morphology changes of AC and AC-TiO$_2$ are shown by the fewer cavities than AC, which suppose that TiO$_2$ occupied the majority of the surface.

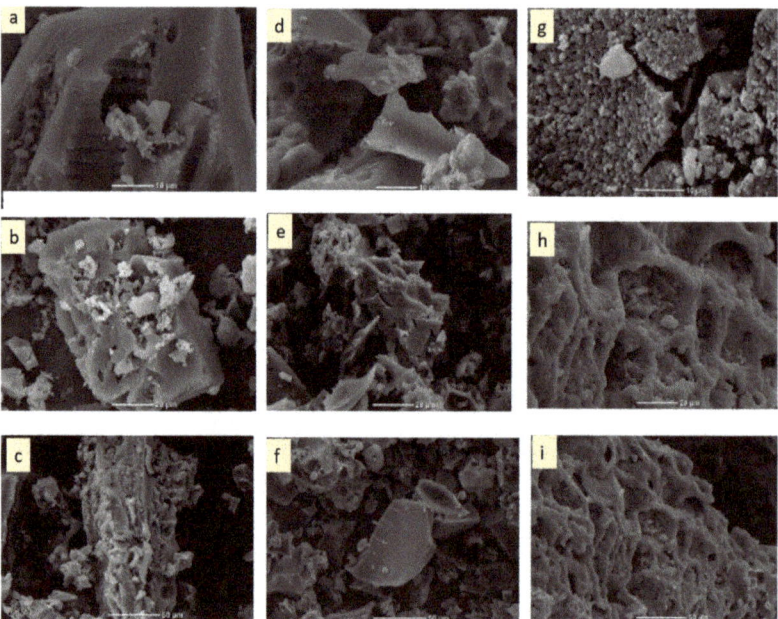

Figure 1. SEM images of AC (**a–c**), AC-OH (**d–f**), and AC-TiO$_2$ (**g–i**).

3.2. Surface Properties

Figures 2 and 3 show the FTIR analysis for all samples to investigate the surface properties and the stability of synthesized samples in liquid media. Starting by measuring the ZP values of all samples, it was found that the ZP of AC-OH is less than 20 mV, suggesting the good stability of the materials. Although, ZP results confirmed the successful addition of hydroxyl groups at the AC surface, as supported by noticeable decreases in ZP. The surface charge of the particles has a potential effect on catalytic activity, as it could involve such interaction with pharmaceutical pollutants. The results on the AC-TiO$_2$

surface displayed a marked decrease in the zeta potential from 8 mV for AC to −14 mV for AC-TiO$_2$. The negative surface charge could play a key role in the adsorption of pollutants from the water.

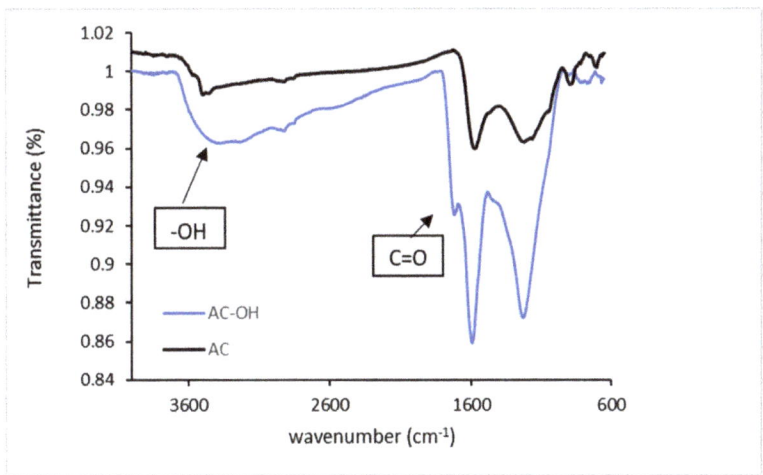

Figure 2. FTIR spectra of AC and AC-OH.

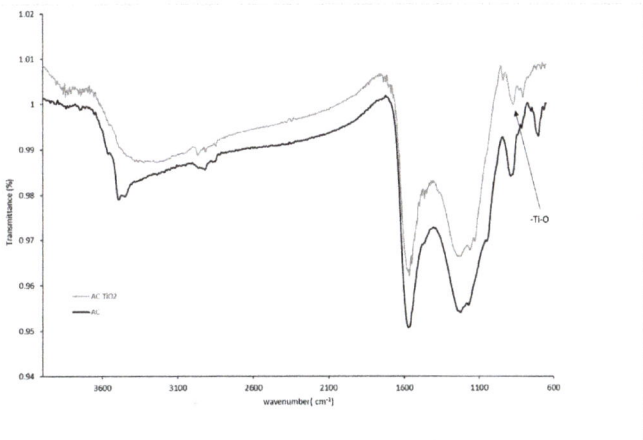

Figure 3. FTIR spectra of AC and AC-TiO$_2$.

According to the Fourier transform infrared spectrum shown in Figure 2, the peaks of the AC surface before and after hydroxylation treatment are different from one sample to another. For AC, the peak located at 1579 cm^{-1} is assigned to the stretching vibration of C=C, whereas the band observed at 1259 cm^{-1} is attributed to the bending vibration of C-H in the methylene group. At the same time, the spectrum of AC-OH displayed a visible change due to the hydroxylation steps. The new band that appeared at 3415 cm^{-1} is associated with hydroxyl groups' stretching vibration (-OH). Accordingly, the chemical structure of the synthesized samples, the bands centered at around 1720 cm^{-1} and 1105 cm^{-1}, are attributed to the presence of the C=O and C-O groups, respectively.

A comparison between the FTIR spectra of AC and AC-TiO$_2$ is shown in Figure 3. The stretching vibration of the hydroxyl group registered at 3400 cm^{-1} is assigned to the physisorbed surface water [34]. A slight shift to a lower wavenumber is observed and explained by the presence of TiO_2 on the AC surface. According to Loo et al., the peak that appeared at 768 cm^{-1} is assigned to the stretching vibration of Ti-O [34].

3.3. Thermal Properties

DSC analysis was performed to investigate the thermal properties of AC and AC-OH. AC samples produced a higher endothermic peak. This peak detected around 90 °C corresponds to the dehydration of the adsorbents. In addition, AC displayed an exothermic peak from 410 to 500 °C associated with the degradation of cellulose, lignin, and hemicellulose. DSC thermograms of AC-OH (Figure 4) display two endothermic peaks around 90 and 400 °C associated with the dehydration and dihydroxylations of the date stem. All of these results correspond to the literature as well as SEM and IR data.

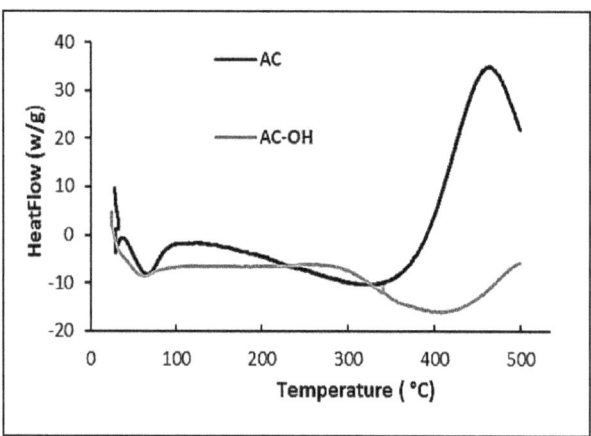

Figure 4. DSC analysis of AC and AC-OH.

3.4. Adsorption of Atenolol and Propranolol

The adsorption of AT and PR was carried out in the presence of AC and AC-OH. A commercial activated carbon (AC) was also evaluated for removing AT and PR from contaminated water. To investigate the parameter effects on the adsorption of pharmaceutical products, temperature, initial concentration, and pH were studied in terms of adsorption capacity. The results are shown in detail in the Supplementary File. As seen in Figures S2–S4, the high temperature was not suitable for the adsorption of AT nor PR molecules due to their limited temperature stability. The adsorption capacity increased as the concentration of pollutants increased to achieve the equilibrium phase, indicating its saturation [14,26,35]. Measurements on the effect of pH displayed that the adsorption of pharmaceutical molecules could have been performant in basic media, while the acidic solution avoided competition with a proton. The removal capacities of the samples, which had the same experimental conditions, i.e., the synthetic and the commercial AC, are reported in Figure 5. The results show that both samples adsorbed the pharmaceutical products in two distinct phases. A rapid adsorption process describes the first phase, while equilibrium steps characterize the second phase. The large number of active sites available at the AC surface for the adsorption of AT and PR explains this result. Meanwhile, over 60 min, the main part of these active sites became saturated by adsorbate, resulting in limited access to more molecules in the solution. Therefore, the second step was reached to achieve the adsorption equilibrium [14].

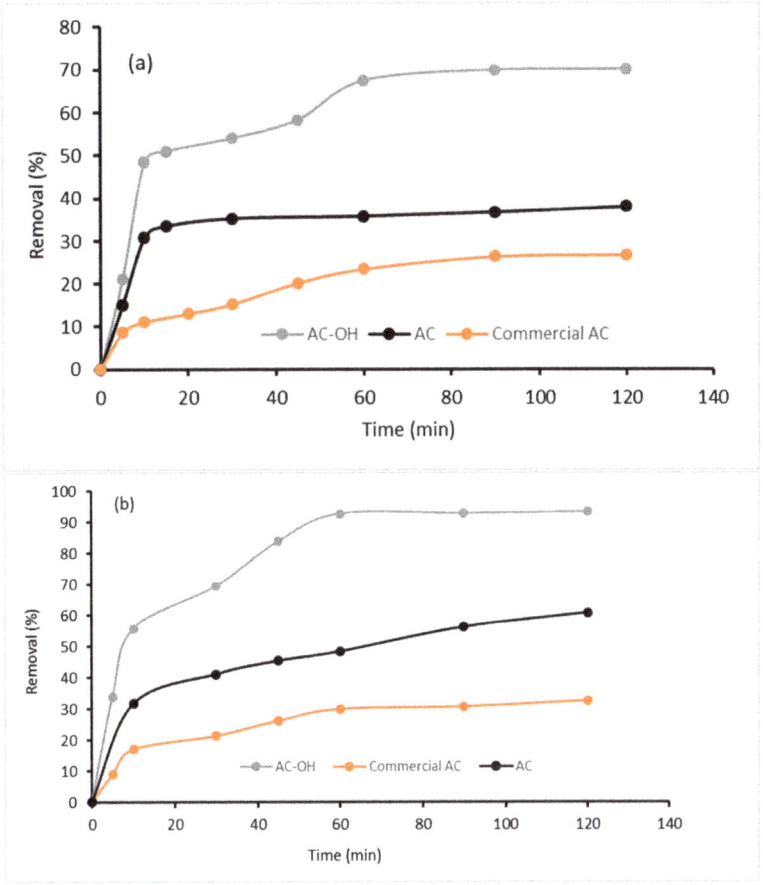

Figure 5. Removal capacity of (**a**) AT and (**b**) PR over AC, AC-OH and commercial AC.

A comparison of the adsorption capacity of commercial and prepared AC shows that AC synthesized from date stems produced a higher adsorption uptake (60%) than commercial AC (32%). This trend can be explained by the key role of treatment and carbonization in this work. It was found that our prepared AC's equilibrium time is longer than the commercial AC's. Interestingly, the best performance is observed for the AC-OH sample, where the adsorption capacity is over 93% for the removal of PR in 120 min. The adsorption kinetic is also faster with AC-OH because the equilibrium was reached after 60 min, and only 10 min is required to adsorb 56% of the PR. This result can be explained by the negative charge surface of AC-OH, as supported by the Zeta potential measurement. For the atenolol molecule, the performance of AC and AC-OH is still higher than that of AC, but the adsorption was more difficult than for propranolol. Indeed, 67% of the AT was adsorbed on AC-OH compared to 36% and 23% for AC (from date stem) and commercial AC, respectively.

3.5. Photocatalytic Degradation of Atenolol and Propranolol

The effect of TiO_2 concentration is considered the first parameter that can affect the removal efficiency of the catalyst. The photocatalytic activity of AC-TiO_2 was measured for various concentrations of TiO_2 (30%, 50%, and 70%); the results are shown in Figure S6. The results show that AT and PR's photocatalytic degradation depends on the TiO_2 quantity.

While the TiO$_2$ concentration increased, photocatalytic degradation increased. The rise of photocatalyst radicals can explain this. The degradation continued until reaching an optimum close to 50%. Measurements on the AC-OH-supported TiO$_2$ were investigated regarding adsorption capacity for both molecules AT and PR. Figure 6 depicts the results obtained for the photocatalytic efficiency of AC-OH/TiO$_2$ and AC-TiO$_2$.

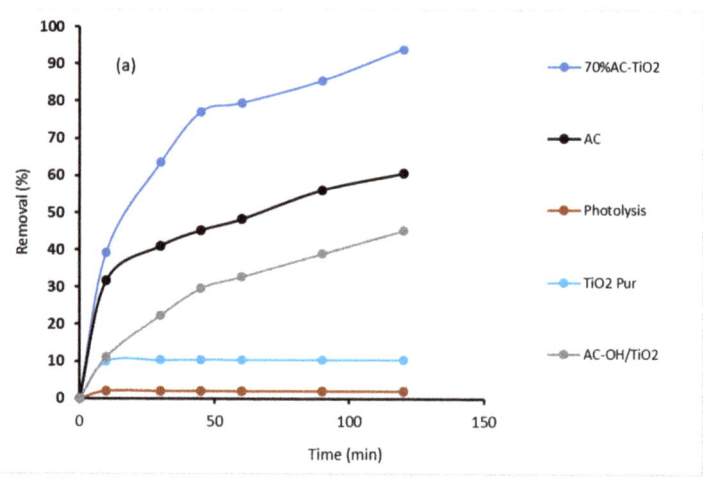

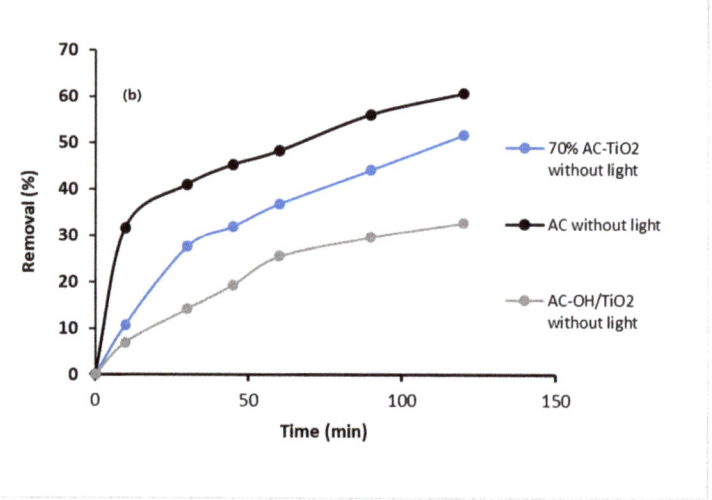

Figure 6. Removal of PR (**a**) with and (**b**) without light using AC, TiO$_2$, AC-OH, AC-OH-/TiO$_2$, and AC-TiO$_2$ catalysts.

According to the obtained results, it is clear that AC-OH/TiO$_2$ can eliminate more than 45% of PR, while AC-TiO$_2$ achieves 94% degradation. Photolysis and pure TiO$_2$ are tested separately to understand the high degradation capacity. Only 10% PR removal is obtained for the materials, suggesting the potential role of the coating process for AC-OH and TiO$_2$.

Experiments were carried out in the presence of and without light to demonstrate the beneficial effect of irradiation light combined with AC-OH-TiO$_2$ and AC-TiO$_2$. Figure 7 demonstrates that light is combined with AC-TiO$_2$ to obtain high adsorption uptake. The removal of PR and AT by AC-OH-TiO$_2$ and AC-TiO$_2$ without light is mainly due to their adsorption on the active surface. Results showed that the photocatalytic degradation efficiency increased significantly for all AC-TiO$_2$ composites compared to pure TiO$_2$. When (PR and AT) are adsorbed on the AC surface, they react with reactive radicals by an oxidation reaction; this improves the catalyst activity. Hayati et al. [36] confirmed that AC is very important because it could act as an electron sink, which allows for the interfacial transfer of photo-induced electrons from TiO$_2$ to AC and the inhibition of the electron recombination rate.

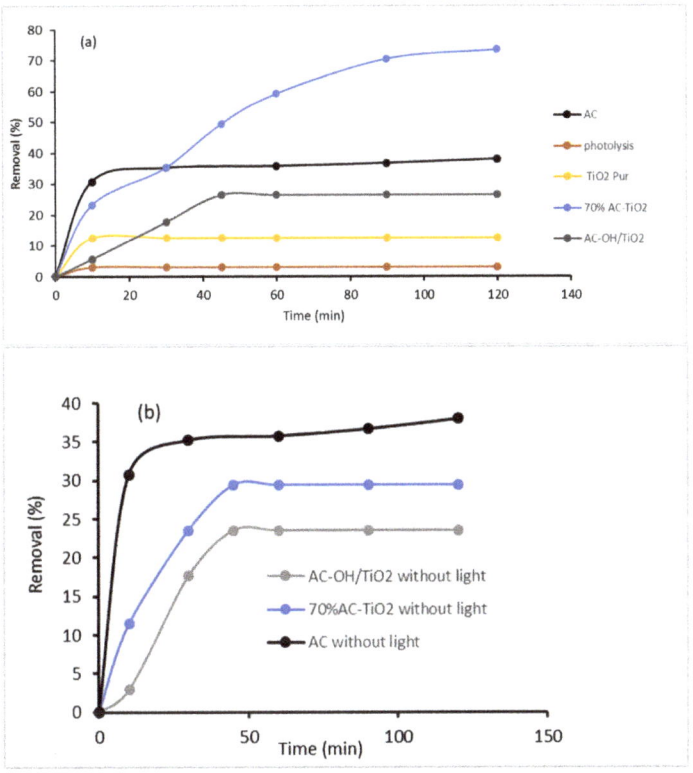

Figure 7. Removal of AT (**a**) with and (**b**) without light using AC, TiO$_2$, AC-OH, AC-OH/TiO$_2$, and AC-TiO$_2$ catalysts.

3.6. Stability and Reusability of the Prepared Materials

The reusability of the prepared materials is investigated and evaluated. The AC-based adsorbent is tested for various adsorption cycles, and the results are displayed in Figure 8.

From the recyclability test, it was established that the system could work with a removal efficiency close to 50% for the studied molecules (AT, 46%; PR, 58%) for a minimum of four cycles. After each adsorption cycle, the adsorbent AC-OH is used for the desorption system, in which the leaching solvent is methanol. The histogram of the removal of AT and PR after each cycle is represented in Figure 8. As seen, for the PR removal, in the first cycle, percent recovery rates of 92% were reached in 2 h, while the second cycle shifted to 87%, which can be considered appreciable results for the recycled adsorbent. The same trend was observed for the removal of AT, which presents a slight decrease from 70% to 65%. Importantly, the developed materials had good recyclability, reaching 52% of PR removal over the fourth cycle. Accordingly, the obtained results could be considered a suitable and stable adsorbent for removing pharmaceutical products.

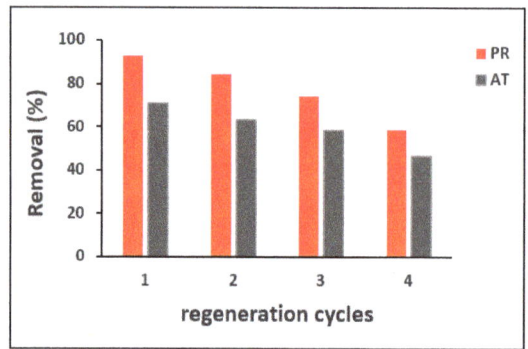

Figure 8. Effect of regeneration cycle on adsorption removal for PR and AT pharmaceutical contaminants.

The adsorption capacities of the pharmaceutical molecules were compared with data already published using other adsorbents. Table 1 summarizes the maximum adsorption capacity over various adsorbents for treating AT and PR. Comparatively, it was found that AC-OH showed a better adsorption capacity for AT (288 mg/g) and PR (339 mg/g) compared to the other adsorbents, such as granular activated carbon, hematite nanoparticles, and activated carbon fiber. Interestingly, this finding can open a new prospect to evaluate the prepared materials for other toxic molecules on a large scale, like the industrial scale.

In addition, it will be important to evaluate the efficiency of AC-TiO$_2$ on the photodegradation uptake. It was clear that AC-TiO$_2$ performs well on the photodegradation of AT and PR in liquid solution. Table 2 reports the photocatalytic degradation capacity using AC-TiO$_2$ compared to other commercial photocatalysts reported in the literature. However, the photocatalytic degradation of pollutants by TiO$_2$ has been extensively studied, while there are few reports of utilizing AC-TiO$_2$ to degrade atenolol and propranolol (Table 2). It can be observed that AC-TiO$_2$ is an efficient candidate to eliminate AT and PR, as compared to the other catalysts. It is interesting to note that AC-TiO$_2$ is efficient in treating concentrated solutions of AT and PR in a short time compared to data already published in the literature. It should also be reminded that AC-TiO$_2$ is inexpensive as it is prepared from a lignocellulosic agro-waste.

3.7. Comparison between the Adsorption and the Photocatalysis Methods

As mentioned in the introduction, this work comprehensively compared two methods for the water treatment, i.e., adsorption and photodegradation techniques, using the same support like AC-OH and AC-TiO$_2$ for adsorption and photodegradation, respectively. Despite the interesting results obtained in the above section for both processes, a distinguishing difference can be illustrated in many aspects, particularly the reusability and the cost. Accordingly, the most useful and more suitable methods for wastewater treatment can be selected. On the one hand, the photocatalytic degradation of pharmaceutical products is based on utilizing TiO$_2$ nanoparticles, which represents one of the most promising methods to decontaminate water containing organic pollutants, such as AT and PR. In its nanometric form, titanium dioxide is considered the first interesting catalyst for photocatalytic wastewater treatments, as it presents high stability. Also, in light of the results obtained in this work, it was found that TiO$_2$ incorporated over AC is attractive due to its highly reactive surfaces, which make them a good photocatalyst of pollutants. Unlike the high energy consumption caused by the UV light during the sorbent activation, nanoparticle powders can be released in treated water, contaminating the environment. As the toxicity of the environment is not accepted to any degree, photodegradation via AC-TiO$_2$ is limited and cannot be developed on a large scale, including any industrialization of the process. However, the adsorption process represents an appreciable removal capacity of up to 89% of AT and PR pharmaceutical pollutants. Likewise, the adsorption methods are widely employed since it provides many advantages, such as low energy consumption, strong reusability, cost-effectiveness, and easy handling.

Table 1. Comparison of the adsorption capacity of AT and PR over various adsorbents.

Pharmaceutical Molecules	Adsorbent	Maximum Adsorption Capacity (mg/g)	References
Atenolol	Waste-biomass-derived activated carbon	183.7	[37]
	Graphene oxide	93.0	[35]
	Granular activated carbon	1.2	[14]
	Activated palm kernel shell	0.2	[37]
	Multiwalled carbon nanotube	5.1	[38]
	Corncob biochar-Mt	90.0	[39]
	AC	119.7	This work
	AC-OH	288.0	This work
Propranolol	Graphene oxide	68.0	[35]
	Activated carbon fiber	0.3	[40]
	Smectite clay mineral montmorillonite	161	[41]
	AC	202	This work
	AC-OH	339.0	This work

Table 2. Comparison between photocatalytic degradation of (AT and PR) in this study with other catalysts reported in the literature.

	Catalyst	Light Source	Concentration (mg/L)	Time (h) to Reach Maximal Degradation	References
Propranolol	Aeroxide TiO$_2$ P25	Low-pressure mercury lamp	26	3 h (92%)	[9]
	AC-TiO$_2$	UV-lamp	50	2 h (94%)	This work
	Degussa P25 TiO$_2$	Xe lamp	15	1 h (100%)	[42]
	Ag-TiO$_2$	High-pressure mercury lamp	20	0.5 h (92%)	[43]
Atenolol	TiO$_2$/Salicylaldehyde-NH$_2$-MIL-101	Xe lamp	10	1 h (82%)	[44]
	Quartz fiber–TiO$_2$	High-pressure mercury lamp	0.002	6 h (50%)	[45]
	Aeroxide TiO$_2$ P25	Low-pressure mercury lamp	80	3 h (87%)	[9]
	AC-TiO$_2$	UV-lamp	50	2 h (73%)	This work

Interestingly, no secondary contamination can be caused by this eco-friendly adsorption during water treatment. To date, wastewater treatment by adsorption can be used in full-scale applications, and very effective results are obtained using this method [15,16,46,47]. On the other, to assess the possible environmental impact and the safety of the whole process, the absence of toxic effects of the process effluents must be guaranteed. Accordingly, the adsorption process is more efficient than photodegradation methods.

4. Conclusions

We have compared the adsorption and the photodegradation methods in wastewater treatment, particularly the removal of pharmaceutical products (AT and PR). In this regard, biomasses based on date stem were activated and calcinated to prepare an activated carbon (AC). The AC materials were modified by a hydroxylation strategy to increase the hydroxyl groups over the AC surface, resulting in AC-OH. In addition, to ensure the photodegradation methods, AC was impregnated into TiO_2 solution to produce AC-TiO_2. The prepared materials were characterized and tested for the removal of AT and PR. The results show that the obtained adsorbent exhibited high adsorption capacity for both molecules. The core of the adsorption mechanism involved interactions such as an electrostatic attraction between adsorbate pharmaceutical molecules and AC-OH adsorbent in the aqueous medium. AC-OH regeneration was studied through four adsorption–desorption cycles and found to have a recovery rate of more than 50% after adsorption. A comparison between the treatment methods proved that adsorption is more suitable for removing pollutants from water, as it presents low energy consumption. The study facilitates the preparation of recyclable and stable adsorbent material for wastewater treatment.

Supplementary Materials: The following supporting information can be downloaded at: https://www.mdpi.com/article/10.3390/app13148074/s1, Table S1: Properties of atenolol and propranolol; Figure S1: Chemical structure of propranolol and atenolol; Figure S2: The effect of temperature on the adsorption capacity of AT and PR; Figure S3: Effect of concentration variation in the adsorption capacity of AT and PR; Figure S4: The effect of pH variation in the removal of (**a**) atenolol and (**b**) propranolol; Table S2: Thermodynamic parameters for Langmuir and Freundlich models; Figure S5: Thermodynamic adsorption of AT and PR on AC-OH; Table S3: Thermodynamic adsorption parameters for AT and PR; Figure S6: Removal capacity of (**a**) PR and (**b**) AT by AC doped with various % TiO_2.

Author Contributions: B.S., P.N.F., J.C. and V.M. were involved in the investigation. B.S., N.B., J.V. and G.L.D. were involved in the writing, supervision, and funding acquisition. F.L.D. was involved in the supervision. All authors have read and agreed to the published version of the manuscript.

Funding: This work was financially supported by the University of Rouen Normandy, INSA Rouen Normandy, the Centre National de la Recherche Scientifique (CNRS), the European Regional Development Fund (ERDF), Labex SynOrg (ANR-11-LABX-0029), the Carnot Institut I2C, the Graduate School for Research Xl-Chem (ANR-18-EURE-0020 XL CHEM), and by the Region Normandie and the Grand Evreux Agglomeration.

Institutional Review Board Statement: Not applicable.

Informed Consent Statement: Not applicable.

Data Availability Statement: Complementary data can be obtained upon request.

Conflicts of Interest: The authors declare no conflict of interest.

References

1. Sangari, N.U.; Jothi, B.; Devi, S.C.; Rajamani, S. Template free synthesis, characterization and application of nano ZnO rods for the photocatalytic decolourization of methyl orange. *J. Water Process Eng.* **2016**, *12*, 1–7. [CrossRef]
2. Singh, N.; Balomajumder, C. Simultaneous removal of phenol and cyanide from aqueous solution by adsorption onto surface modified activated carbon prepared from coconut shell. *J. Water Process Eng.* **2016**, *9*, 233–245. [CrossRef]
3. Yang, Q.; Liao, Y.; Mao, L. Kinetics of Photocatalytic Degradation of Gaseous Organic Compounds on Modified TiO_2/AC Composite Photocatalyst. *Chin. J. Chem. Eng.* **2012**, *20*, 572–576. [CrossRef]
4. Daughton, C.G.; Ternes, T.A. Pharmaceuticals and personal care products in the environment: Agents of subtle change? *Environ. Health Perspect.* **1999**, *107* (Suppl. S6), 907–938. [CrossRef]
5. Boxall, A.B.; Rudd, M.A.; Brooks, B.W.; Caldwell, D.J.; Choi, K.; Hickmann, S.; Innes, E.; Ostapyk, K.; Staveley, J.P.; Verslycke, T.; et al. Pharmaceuticals and personal care products in the environment: What are the big questions? *Environ. Health Perspect.* **2012**, *120*, 1221–1229. [CrossRef]
6. Sirés, I.; Brillas, E. Remediation of water pollution caused by pharmaceutical residues based on electrochemical separation and degradation technologies: A review. *Environ. Int.* **2012**, *40*, 212–229. [CrossRef]
7. Boyd, G.R.; Reemtsma, H.; Grimm, D.A.; Mitra, S. Pharmaceuticals and personal care products (PPCPs) in surface and treated waters of Louisiana, USA and Ontario, Canada. *Sci. Total Environ.* **2003**, *311*, 135–149. [CrossRef]
8. Ducey, S.B.; Sapkota, A. *Presence of Pharmaceuticals and Personal Care Products in the Environment—A Concern for Human Health?* ACS Symposium Series; Oxford University Press: Oxford, UK, 2010; pp. 345–365.
9. Ponkshe, A.; Thakur, P. Significant mineralization of beta blockers Propranolol and Atenolol by TiO_2 induced photocatalysis. *Mater. Today Proc.* **2019**, *18*, 1162–1175. [CrossRef]

10. Seema, M.D. In vitro sustained delivery of atenolol, an antihypertensive drug using naturally occurring clay mineral montmorillonite as a carrier. *Eur. Chem. Bull.* **2013**, *2*, 941–951.
11. Dong, M.M.; Trenholm, R.; Rosario-Ortiz, F.L. Photochemical degradation of atenolol, carbamazepine, meprobamate, phenytoin and primidone in wastewater effluents. *J. Hazard. Mater.* **2015**, *282*, 216–223. [CrossRef]
12. Küster, A.; Alder, A.C.; Escher, B.I.; Duis, K.; Fenner, K.; Garric, J.; Hutchinson, T.H.; Lapen, D.R.; Péry, A.; Römbke, J.; et al. Environmental risk assessment of human pharmaceuticals in the European Union: A case study with the β-blocker atenolol. *Integr. Environ. Assess. Manag.* **2010**, *6*, 514–523. [CrossRef] [PubMed]
13. Li, Z.; Fitzgerald, N.M.; Lv, G.; Jiang, W.-T.; Wu, L. Adsorption of Atenolol on Talc: An Indication of Drug Interference with an Excipient. *Adsorpt. Sci. Technol.* **2015**, *33*, 379–392. [CrossRef]
14. Haro, N.K.; Del Vecchio, P.; Marcilio, N.R.; Féris, L.A. Removal of atenolol by adsorption—Study of kinetics and equilibrium. *J. Clean. Prod.* **2017**, *154*, 214–219. [CrossRef]
15. Bouazizi, N.; Vieillard, J.; Samir, B.; Le Derf, F. Advances in Amine-Surface Functionalization of Inorganic Adsorbents for Water Treatment and Antimicrobial Activities: A Review. *Polymers* **2022**, *14*, 378. [CrossRef]
16. Nkuigue Fotsing, P.; Bouazizi, N.; Djoufac Woumfo, E.; Mofaddel, N.; Le Derf, F.; Vieillard, J. Investigation of chromate and nitrate removal by adsorption at the surface of an amine-modified cocoa shell adsorbent. *J. Environ. Chem. Eng.* **2021**, *9*, 104618. [CrossRef]
17. Vieillard, J.; Bouazizi, N.; Morshed, M.N.; Clamens, T.; Desriac, F.; Bargougui, R.; Thebault, P.; Lesouhaitier, O.; Le Derf, F.; Azzouz, A. CuO Nanosheets Modified with Amine and Thiol Grafting for High Catalytic and Antibacterial Activities. *Ind. Eng. Chem. Res.* **2019**, *58*, 10179–10189. [CrossRef]
18. Bouazizi, N.; Vieillard, J.; Bargougui, R.; Couvrat, N.; Thoumire, O.; Morin, S.; Ladam, G.; Mofaddel, N.; Brun, N.; Azzouz, A.; et al. Entrapment and stabilization of iron nanoparticles within APTES modified graphene oxide sheets for catalytic activity improvement. *J. Alloys Compd.* **2019**, *771*, 1090–1102. [CrossRef]
19. Bouazizi, N.; Vieillard, J.; Thebault, P.; Desriac, F.; Clamens, T.; Bargougui, R.; Couvrat, N.; Thoumire, O.; Brun, N.; Ladam, G.; et al. Silver nanoparticle embedded copper oxide as an efficient core–shell for the catalytic reduction of 4-nitrophenol and antibacterial activity improvement. *Dalton Trans.* **2018**, *47*, 9143–9155. [CrossRef]
20. Yap, P.-S.; Lim, T.-T.; Srinivasan, M. Nitrogen-doped TiO2/AC bi-functional composite prepared by two-stage calcination for enhanced synergistic removal of hydrophobic pollutant using solar irradiation. *Catal. Today* **2011**, *161*, 46–52. [CrossRef]
21. Sampaio, M.J.; Silva, C.G.; Silva, A.M.T.; Vilar, V.J.P.; Boaventura, R.A.R.; Faria, J.L. Photocatalytic activity of TiO2-coated glass raschig rings on the degradation of phenolic derivatives under simulated solar light irradiation. *Chem. Eng. J.* **2013**, *224*, 32–38. [CrossRef]
22. Huang, D.; Miyamoto, Y.; Matsumoto, T.; Tojo, T.; Fan, T.; Ding, J.; Guo, Q.; Zhang, D. Preparation and characterization of high-surface-area TiO2/activated carbon by low-temperature impregnation. *Sep. Purif. Technol.* **2011**, *78*, 9–15. [CrossRef]
23. Aruldoss, U.; Kennedy, L.J.; Judith Vijaya, J.; Sekaran, G. Photocatalytic degradation of phenolic syntan using TiO2 impregnated activated carbon. *J. Colloid Interface Sci.* **2011**, *355*, 204–209. [CrossRef]
24. Zhu, C.; Wang, X.; Huang, Q.; Huang, L.; Xie, J.; Qing, C.; Chen, T. Removal of gaseous carbon bisulfide using dielectric barrier discharge plasmas combined with TiO$_2$ coated attapulgite catalyst. *Chem. Eng. J.* **2013**, *225*, 567–573. [CrossRef]
25. Hinojosa-Reyes, M.; Arriaga, S.; Diaz-Torres, L.A.; Rodríguez-González, V. Gas-phase photocatalytic decomposition of ethylbenzene over perlite granules coated with indium doped TiO$_2$. *Chem. Eng. J.* **2013**, *224*, 106–113. [CrossRef]
26. Zhang, X.; Lei, L. Effect of preparation methods on the structure and catalytic performance of TiO$_2$/AC photocatalysts. *J. Hazard. Mater.* **2008**, *153*, 827–833. [CrossRef] [PubMed]
27. Wang, X.; Liu, Y.; Hu, Z.; Chen, Y.; Liu, W.; Zhao, G. Degradation of methyl orange by composite photocatalysts nano-TiO$_2$ immobilized on activated carbons of different porosities. *J. Hazard. Mater.* **2009**, *169*, 1061–1067. [CrossRef]
28. Matos, J.; Garcia, A.; Cordero, T.; Chovelon, J.-M.; Ferronato, C. Eco-friendly TiO$_2$–AC Photocatalyst for the Selective Photooxidation of 4-Chlorophenol. *Catal. Lett.* **2009**, *130*, 568–574. [CrossRef]
29. Gomathi Devi, L.; Narasimha Murthy, B. Characterization of Mo Doped TiO$_2$ and its Enhanced Photo Catalytic Activity Under Visible Light. *Catal. Lett.* **2008**, *125*, 320–330. [CrossRef]
30. Jamil, T.S.; Ghaly, M.Y.; Fathy, N.A.; Abd El-Halim, T.A.; Österlund, L. Enhancement of TiO$_2$ behavior on photocatalytic oxidation of MO dye using TiO$_2$/AC under visible irradiation and sunlight radiation. *Sep. Purif. Technol.* **2012**, *98*, 270–279. [CrossRef]
31. Gar Alalm, M.; Tawfik, A.; Shinichi, O. Combined Solar advanced oxidation and PAC adsorption for removal of pesticides from industrial wastewater. *J. Mater. Environ. Sci.* **2015**, *6*, 800–809.
32. Bakhta, S.; Sadaoui, Z.; Bouazizi, N.; Samir, B.; Allalou, O.; Devouge-Boyer, C.; Mignot, M.; Vieillard, J. Functional activated carbon: From synthesis to groundwater fluoride removal. *RSC Adv.* **2022**, *12*, 2332–2348. [CrossRef] [PubMed]
33. Samir, B.; Bakhta, S.; Bouazizi, N.; Sadaoui, Z.; Allalou, O.; Le Derf, F.; Vieillard, J. TBO Degradation by Heterogeneous Fenton-like Reaction Using Fe Supported over Activated Carbon. *Catalysts* **2021**, *11*, 1456. [CrossRef]
34. Rashid, M.M.; Zorc, M.; Simončič, B.; Jerman, I.; Tomšič, B. In-Situ Functionalization of Cotton Fabric by TiO2: The Influence of Application Routes. *Catalysts* **2022**, *12*, 1330. [CrossRef]
35. Hayati, F.; Khodabakhshi, M.R.; Isari, A.A.; Moradi, S.; Kakavandi, B. LED-assisted sonocatalysis of sulfathiazole and pharmaceutical wastewater using N,Fe co-doped TiO$_2$@SWCNT: Optimization, performance and reaction mechanism studies. *J. Water Process Eng.* **2020**, *38*, 101693. [CrossRef]

36. To, M.-H.; Hadi, P.; Hui, C.-W.; Lin, C.S.K.; McKay, G. Mechanistic study of atenolol, acebutolol and carbamazepine adsorption on waste biomass derived activated carbon. *J. Mol. Liq.* **2017**, *241*, 386–398. [CrossRef]
37. Kyzas, G.Z.; Koltsakidou, A.; Nanaki, S.G.; Bikiaris, D.N.; Lambropoulou, D.A. Removal of beta-blockers from aqueous media by adsorption onto graphene oxide. *Sci. Total Environ.* **2015**, *537*, 411–420. [CrossRef]
38. Dehdashti, B.; Amin, M.M.; Pourzamani, H.; Rafati, L.; Mokhtari, M. Removal of atenolol from aqueous solutions by multiwalled carbon nanotubes modified with ozone: Kinetic and equilibrium study. *Water Sci. Technol. J. Int. Assoc. Water Pollut. Res.* **2018**, *2017*, 636–649. [CrossRef]
39. Fu, C.; Zhang, H.; Xia, M.; Lei, W.; Wang, F. The single/co-adsorption characteristics and microscopic adsorption mechanism of biochar-montmorillonite composite adsorbent for pharmaceutical emerging organic contaminant atenolol and lead ions. *Ecotoxicol. Environ. Saf.* **2020**, *187*, 109763. [CrossRef]
40. Zhao, Y.; Cho, C.-W.; Wang, D.; Choi, J.-W.; Lin, S.; Yun, Y.-S. Simultaneous scavenging of persistent pharmaceuticals with different charges by activated carbon fiber from aqueous environments. *Chemosphere* **2020**, *247*, 125909. [CrossRef]
41. Del Mar Orta, M.; Martín, J.; Medina-Carrasco, S.; Santos, J.L.; Aparicio, I.; Alonso, E. Adsorption of propranolol onto montmorillonite: Kinetic, isotherm and pH studies. *Appl. Clay Sci.* **2019**, *173*, 107–114. [CrossRef]
42. Medana, C.; Calza, P.; Carbone, F.; Pelizzetti, E.; Hidaka, H.; Baiocchi, C. Characterization of atenolol transformation products on light-activated TiO_2 surface by high-performance liquid chromatography/high-resolution mass spectrometry. *Rapid Commun. Mass Spectrom. RCM* **2008**, *22*, 301–313. [CrossRef] [PubMed]
43. Ling, Y.; Liao, G.; Xie, Y.; Yin, J.; Huang, J.; Feng, W.; Li, L. Coupling photocatalysis with ozonation for enhanced degradation of Atenolol by Ag-TiO_2 micro-tube. *J. Photochem. Photobiol. A Chem.* **2016**, *329*, 280–286. [CrossRef]
44. Mehrabadi, Z.; Faghihian, H. Comparative photocatalytic performance of TiO_2 supported on clinoptilolite and TiO_2/Salicylaldehyde-NH_2-MIL-101(Cr) for degradation of pharmaceutical pollutant atenolol under UV and visible irradiations. *J. Photochem. Photobiol. A Chem.* **2018**, *356*, 102–111. [CrossRef]
45. Arlos, M.J.; Hatat-Fraile, M.M.; Liang, R.; Bragg, L.M.; Zhou, N.Y.; Andrews, S.A.; Servos, M.R. Photocatalytic decomposition of organic micropollutants using immobilized TiO_2 having different isoelectric points. *Water Res.* **2016**, *101*, 351–361. [CrossRef] [PubMed]
46. Morshed, M.N.; Bouazizi, N.; Behary, N.; Vieillard, J.; Thoumire, O.; Nierstrasz, V.; Azzouz, A. Iron-loaded amine/thiol functionalized polyester fibers with high catalytic activities: A comparative study. *Dalton Trans.* **2019**, *48*, 8384–8399. [CrossRef] [PubMed]
47. Nabil, B.; Ahmida, E.-A.; Christine, C.; Julien, V.; Abdelkrim, A. Inorganic-organic-fabrics based polyester/cotton for catalytic reduction of 4-nitrophenol. *J. Mol. Struct.* **2019**, *1180*, 523–531. [CrossRef]

Disclaimer/Publisher's Note: The statements, opinions and data contained in all publications are solely those of the individual author(s) and contributor(s) and not of MDPI and/or the editor(s). MDPI and/or the editor(s) disclaim responsibility for any injury to people or property resulting from any ideas, methods, instructions or products referred to in the content.

Article

Dose–Response Effect of Nitrogen on Microbial Community during Hydrocarbon Biodegradation in Simplified Model System

Justyna Staninska-Pięta [1,*], Jakub Czarny [2], Wojciech Juzwa [3], Łukasz Wolko [4], Paweł Cyplik [3,*] and Agnieszka Piotrowska-Cyplik [1,*]

1. Department of Food Technology of Plant Origin, Poznan University of Life Sciences, Wojska Polskiego 31, 60-624 Poznan, Poland
2. Institute of Forensic Genetics, Al. Mickiewicza 3/4, 85-071 Bydgoszcz, Poland; pubjc@igs.org.pl
3. Department Biotechnology and Food Microbiology, Poznan University of Life Sciences, Wojska Polskiego 48, 60-627 Poznan, Poland; wojciech.juzwa@up.poznan.pl
4. Department of Biochemistry and Biotechnology, Poznan University of Life Sciences, Dojazd 11, 60-632 Poznan, Poland; lukasz.wolko@up.poznan.pl
* Correspondence: justyna.staninska@up.poznan.pl (J.S.-P.); pawel.cyplik@up.poznan.pl (P.C.); agnieszka.piotrowska-cyplik@up.poznan.pl (A.P.-C.)

Abstract: Knowledge about the influence of C:N ratio on the biodegradation process of hydrocarbon compounds is of significant importance in the development of biostimulation techniques. The purpose of this study was to assess the impact of nitrogen compounds on the environmental consortium during the process of biological decomposition of hydrocarbons. The experimental variants represented low, moderate, and excessive biostimulation with nitrogen compounds. The metabolic activity of the consortium was tested using the flow cytometry technique. The efficiency of the biodegradation of hydrocarbons of the consortium, based on the gas chromatography method, and metapopulation changes, based on the analysis of V4 16srRNA sequencing data, were assessed. The results of the research confirm the positive effect of properly optimized biostimulation with nitrogen compounds on the biological decomposition of polycyclic aromatic hydrocarbons. The negative impact of excessive biostimulation on the biodegradation efficiency and metabolic activity of microorganisms is also proven. Low resistance to changes in the supply of nitrogen compounds is demonstrated among the orders Xanthomonadales, Burkholderiales, Sphingomonadales, Flavobacteriales, and Sphingobacteriales. It is proven that quantitative analysis of the order of Rhizobiales, characterized by a high-predicted potential for the decomposition of polycyclic aromatic hydrocarbons, may be helpful during biostimulation optimization processes in areas with a high nitrogen deficiency.

Keywords: microbial community; hydrocarbon biodegradation; biostimulation; next generation sequencing

Citation: Staninska-Pięta, J.; Czarny, J.; Juzwa, W.; Wolko, Ł.; Cyplik, P.; Piotrowska-Cyplik, A. Dose–Response Effect of Nitrogen on Microbial Community during Hydrocarbon Biodegradation in Simplified Model System. *Appl. Sci.* **2022**, *12*, 6012. https://doi.org/10.3390/app12126012

Academic Editors: Amanda Laca Pérez and Yolanda Patiño

Received: 18 May 2022
Accepted: 10 June 2022
Published: 13 June 2022

Copyright: © 2022 by the authors. Licensee MDPI, Basel, Switzerland. This article is an open access article distributed under the terms and conditions of the Creative Commons Attribution (CC BY) license (https://creativecommons.org/licenses/by/4.0/).

1. Introduction

Intensive exploitation of liquid fossil fuels, and the migration of hydrocarbon compounds to the natural environment, have a highly negative influence on terrestrial and aquatic ecosystems. Due to the high resistance to biodegradation, and the potential for biomagnification, this pollution is considered to be one of the most serious environmental threats [1–3]. Due to the complexity of the chemical structure, and the diversity of physical and chemical properties, individual groups of hydrocarbons differ in bioavailability, toxicity, and the potential for biological decomposition [3,4].

Many biotic and abiotic factors affect the biological decomposition rate of hydrocarbon compounds [3–8]. The abiotic factor, i.e., the supply of microelements such as nitrogen, phosphorus, sulfur, calcium, magnesium, and potassium, is crucial in the functioning of the cellular metabolism of microorganisms that perform biodegradation processes [9]. In contaminated areas, deficiencies in biogenic elements and inhibition of bioremediation processes may occur very quickly, as a result of an excessive supply of carbon compounds, and the activity of microorganisms [10]. Research suggests that the share of nutrients, including the C:N ratio, should be at a level similar to that in live cells. The optimal values for the C:N ratio are in the range from 100:5 to 100:15 [9,11]. The wide range of the optimal ratio of these elements may result from the different rates of decomposition of individual hydrocarbon groups, and the limited bioavailability of biogenic elements, depending on abiotic and biotic environmental factors [9].

One of the most popular bioremediation technologies, based on the knowledge of the influence of nutrient supply, is the biostimulation technique. Targeted compensation of deficiencies in the environmental niche results in metabolic activation of indigenous microflora, and the improvement of the biodegradation efficiency of xenobiotics. This compensation may consist of the application of deficient nutrients and electron acceptors, in the form of both organic and inorganic compounds [12,13]. Many studies prove that biostimulation is beneficial for effective hydrocarbon degradation [14–16]. A popular strategy is also to combine bioaugmentation and biostimulation techniques, in order to maximize remediation effects [12,17,18].

Despite the many advantages of bioremediation technologies, and their widespread social acceptability, they have a number of limitations. These include the differentiation of the metabolic potential of microorganisms living in different environments, problems related to the optimization of the scale of necessary implementations, and the difficulties in extrapolating the results of tests carried out on a laboratory scale [19,20]. It is emphasized that, in the case of biostimulation with nitrogen compounds, the excessive supply of nutrient may negatively affect the efficiency of biological decomposition [21–23].

An innovative context in the improvement and understanding of the mechanisms of bioremediation processes is the approach to the issues within the field of synthetic microbiology, assuming a multitude of biotic and abiotic interactions in the microbial community. According to the assumptions of the synthetic approach, simple interactions between genotypically different microorganisms may contribute to the creation of properties that are difficult to predict at the consortium level [20,24,25]. Therefore, there is a need to acquire and analyze data on the molecular mechanisms of the microbial transformation of hydrocarbon compounds, and the influence of environmental factors and degradation processes. The aim of this study was to comprehensively assess the effect of nitrogen compounds on an environmental consortium with high hydrocarbon degradation potential. Both the enzymatic efficiency and metabolic activity, as well as the overall genetic potential of individual members of the metapopulation, were taken into account. This type of analysis allows for the improvement of planning an effective and predictable ecosystem remediation process.

2. Materials and Methods

2.1. Microbial Consortium Isolation

The environmental consortium with a high potential for biodegradation of hydrocarbons was isolated from soil matrix. The soil sample was taken from the area of railway sleepers impregnation plant in Solec Kujawski (Poland): 53°04′40.7″ N 18°14′23.6″ E. A detailed description of the soil sampling site is presented in previous publications [8,26].

The isolation protocol was based on our prior experience [8]. A 10 g sample of the soil was added into 90 mL of sterile saline, and shaken for 4 h (150 rpm). After the soil particles sedimented, the obtained supernatant was added to a non-selective nutrient broth, containing enriched broth (BTL, Łódź, Poland) and a 2% of glucose (Sigma-Aldrich, Darmstadt, Germany). Bacteria were incubated for 7 days at 25 °C in aerobic conditions provided by continuous shaking (150 rpm). Finally, the biomass was centrifuged (10 min, 4000 rpm), washed twice, and suspended in saline solution. The microorganism suspension was normalized to OD600 = 0.7 (Helios Delta Vis, ThermoFisher Scientific, Waltham, MA, USA), and used as inoculum for the biodegradation experiments.

2.2. Experimental Design

The biodegradation experiment was performed in flasks equipped with septum and vent caps with a semi-permeable DURAN® membrane, in order to ensure the best oxygenation of the culture medium. Each experimental flask included 50 mL of mineral medium described in our previous studies [8], 2 g of diesel oil (PKN Orlen, Płock, Poland), and 250 µL of optimized microbiological inoculum. Casein peptone (Merck, Darmstadt, Germany) was used as the nitrogen source (Table 1). According to the manufacturer's declarations, the total amount of nitrogen was 14%, of which 4% was the amine nitrogen fraction. Based on the literature data, the nitrogen content in diesel fuel was assumed to be 0.13% [27], and the carbon to nitrogen ratio was assumed to be as 10:1 [11]. Biodegradation was carried out for 168 h at 25 °C. The aerobic conditions were provided by continuous shaking (150 rpm). The experiment was performed in three replicates.

Table 1. Characteristics of the experimental variants.

Designation	Estimated C:N Ratio	Experimental Variant Description
NP0		No nitrogen supplementation (control sample)
NP1	1:1	Low nitrogen supplementation
NP2	10:1	Moderate nitrogen supplementation
NP3	30:1	Excessive nitrogen supplementation

2.3. Hydrocarbon Biodegradation Analysis

The assessment of the loss of individual groups of hydrocarbons was performed after 24, 72, and 168 h of biodegradation, and expressed as a share of the concentration of hydrocarbons directly after inoculation. Gas chromatography, coupled with mass spectrometry (GC–MS), was used for the measurements. The detailed method of extracts preparation and analysis conditions are described in a previous publication [8].

2.4. Metabolic Activity Analysis

In order to evaluate the metabolic population during the biodegradation process, expressed as the redox potential of microbial cells, the method of flow cytometry was used. The bacteria samples were taken after 24 h and 168 h of biodegradation, and stained using BacLight™ RedoxSensor™ Green Vitality Kit (Thermo Fisher Scientific, Waltham, MA, USA). The negative control sample consisted of metabolically inactive, thermally inactivated dead bacteria cells. The analysis protocol was based on the manufacturer instructions, and described in detail in our previous publication [26]. The percentage of the population with high and low metabolic activity was calculated by gating the dot plots of the median fluorescein isothiocyanate fluorescence intensity signals (FITC-A) versus the median signals of the side scatter parameter (SSC-A). The following groups were set: metabolically inactive cells (Q1), metabolically active cells (Q2), and artifacts (Q3 and Q4).

2.5. Genetic Analysis of Microbial Population

2.5.1. DNA Isolation

The isolation of genomic DNA was performed using the Genomic Mini AX Bacteria Spin kit (A&A Biotechnology, Gdańsk, Poland). All isolation steps were conducted according to the protocol provided by the manufacturer. The samples were taken after 168 h of the biodegradation process. The isolation efficiency was controlled by the fluorimetric method on the Qbit 3.0 device, using the Qubit™ dsDNA HS Assay Kit (ThermoFisher Scientific, Waltham, MA, USA). For each sample, three DNA extractions were carried out. Lastly, samples were mixed together, after a positive quantification.

2.5.2. NGS Sequencing

For microbial population taxonomy analysis, the V4 region of the 16SrRNA was amplified, based on the 515F-806R primers designed by Caporaso et al. (2012) [28]. The PCR details were optimized, and described in a previous publication [8].

Sequencing of the obtained amplicons was performed on the MiSeq platform (Illumina, CA, USA). The construction of amplicons and libraries normalization protocol is described in a previous work [26].

2.5.3. Bioinformatic Analysis

The output sequencing data were analyzed using the CLC Genomics Workbench 8.5, with the CLC Microbial Genomics Module 1.2 software (Qiagen, Hilden, Germany). Readings were trimmed, demultiplexed, and paired ends were joined. Then, the chimeric readings were identified and removed. The data were clustered independently against two reference databases at 97% similarity of operational taxonomic units (OTU). The SILVA v119 database [29] was used for taxonomy annotation and biodiversity analysis, and the GreenGenes 13.5 database [30] was used for PICRUSt analysis. Selected biodiversity indices were determined: OTU number, Simpson's index, and phylogenetic diversity. Raw sequence data were deposited in the Sequence Read Archive (SRA) as project no. PRJNA831882.

In order to better characterize the analyzed microbiome, a linear discriminant analysis (LDA) effect size (LEfSe) [31] was performed. It allowed for the identification and selection of the most differentially abundant taxa. The microbial biomarkers that best describe the differences between the groups differing in the level of supplementation with nitrogen compounds were assessed. The following groups were determined: deficiency of nitrogen compounds: variants NP0 and NP1; and no deficiency (sufficient level: variants NP2 and NP3). LDA was coupled with effect size measurements: a non-parametric Kruskal–Wallis rank-sum test ($p < 0.05$), and the Wilcoxon rank-sum test ($p < 0.05$). A threshold of 3.5 for the logarithmic LDA score for discriminative attributes was established as the cut-off value.

2.5.4. PICRUSt Analysis

In order to evaluate the bacterial functional composition in the analyzed samples, the bioinformatic tool PICRUSt v. 1.1.1 was used. This algorithm is widely used in environmental studies to analyze functional assessment of bacteria communities in different environments (soil, sediments, wastewater etc.) [32–38]. The predictions of functional composition of metagenomes were performed on the basis of sequencing data clustered against GreenGenes 13.5, and reference bacterial genomes deposited in the IMG database. The detailed algorithm's mode of action is described in a publication of its authors [39]. Based on the analysis, data on the predicted prevalence of 6009 KEGG Orthology IDs (KO IDs) participating in the degradation of polycyclic, aromatic hydrocarbons were obtained. It was assessed which OTU contributed to particular functions. Quality control of the PICRUSt prediction was performed according to the algorithm authors advices. The genome coverage was calculated for all samples using a weighted-Nearest Sequenced Taxon Index (NSTI) score [39].

2.6. Statistical Analysis

Statistical analysis of the results were accomplished using Statistica v 13.0 software (StatSoft, Kraków, Poland). The processed results were presented in the graphical form as the mean value and the standard deviation. To achieve the objective of verification the hypothesis, the non-parametric tests were applied: Kruskal–Wallis test ($\alpha = 0.05$), and Mann–Whitney test ($\alpha = 0.05$).

3. Results

3.1. Hydrocarbon Biodegradation

The analysis of the biodegradation kinetics shows no significant influence of low and moderate levels of biostimulation (NP1 and NP2) on the distribution of most hydrocarbon fractions: total petroleum hydrocarbons (TPH), alkanes, aromatic, and polyaromatic hydrocarbon compounds (Figure 1). The exception is the fraction of polycyclic aromatic hydrocarbons (PAH), where a statistically significant improvement in the effectiveness of biological decomposition is found in the variants with low and moderate levels of nitrogen supplementation (by 10.8% in the NP1 variant, and 15.2% in the NP2 variant, respectively) (Figure 1D). Among all the analyzed groups of hydrocarbon compounds, the inhibitory effect of excessive supplementation with nitrogen compounds is noted. The strongest biodegradation inhibition is noted in the case of PAH, with only 13.9% of this fraction of hydrocarbon compounds degraded after 7 days.

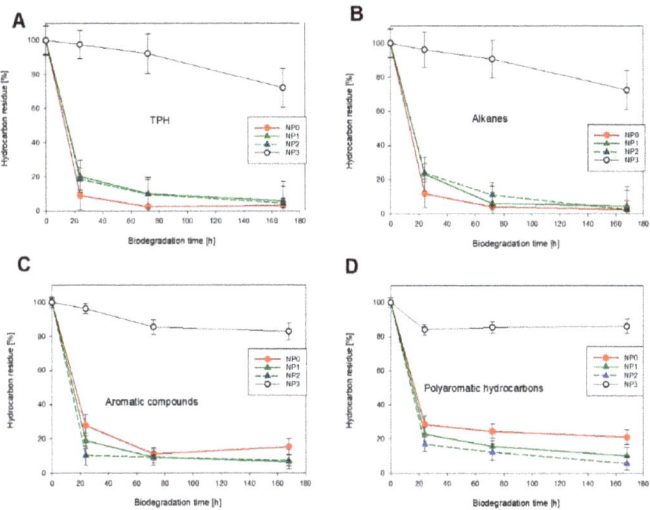

Figure 1. Biodegradation kinetics of selected hydrocarbon fractions: TPH (**A**), alkanes (**B**), aromatic compounds (**C**), and PAH (**D**) in the experimental variants.

3.2. Microbial Activity

Cytometric analysis shows no significant effect of low and moderate doses of nitrogen on the metabolic activity of microbial cells after 24 h of the biodegradation process. In the variant NP3 (excessive biostimulation), a 5.6% decrease in metabolic activity is noted (Figure 2).

The second assessment of metabolic activity, carried out after 168 h of the experiment, shows significant, but slight, changes in the variants NP2 and NP3 (Figure 3). The moderate level of biostimulation (NP2) has a positive effect on the activity of microorganisms (3.2% increase, compared to the control sample). Moreover, excessive biostimulation (NP3) causes a significant decrease in microbial activity (by 17.6%, compared to the control sample) (Figure 3).

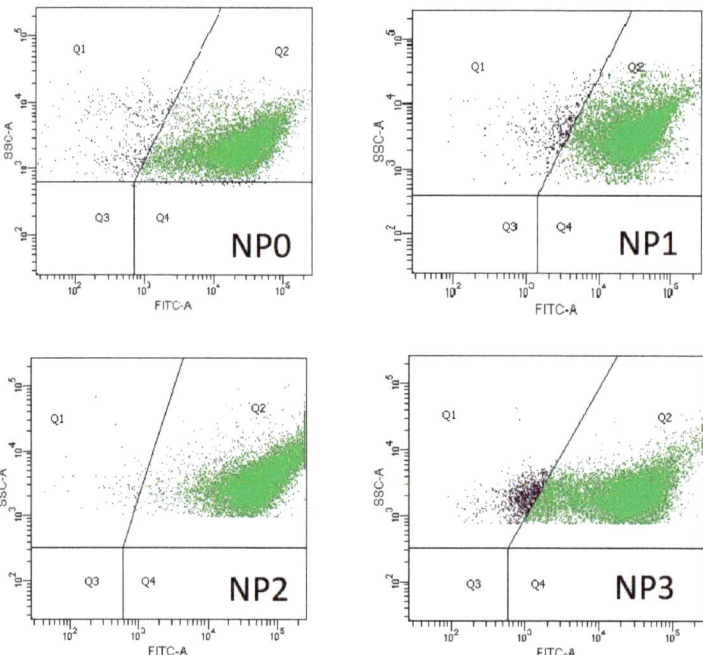

Figure 2. Cytometric analysis of microbial metabolic activity in experimental variants after 24 h of biodegradation. The metabolically active population (Q2) is marked in green, the metabolically inactive population (Q1) is marked in purple.

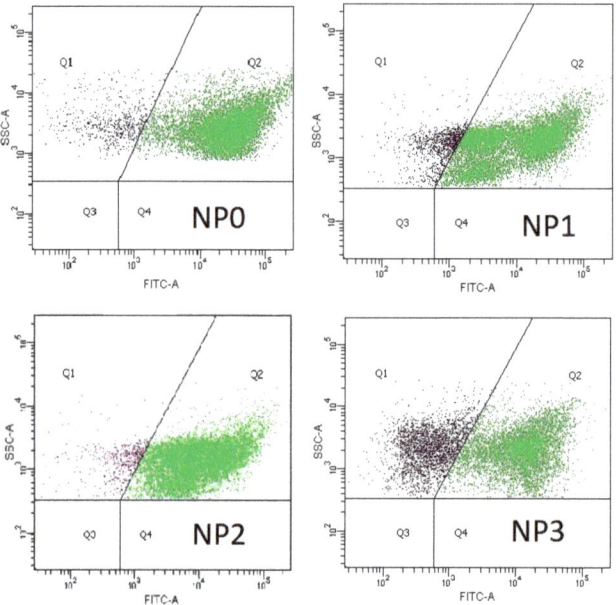

Figure 3. Cytometric analysis of microbial metabolic activity in experimental variants after 168 h of biodegradation. The metabolically active population (Q2) is marked in green, the metabolically inactive population (Q1) is marked in purple.

3.3. Biodiversity Analysis

The results of the selected biodiversity coefficients analysis are presented in Table 2. Excessive levels of biostimulation (NP3) contribute to a significant decrease in biodiversity (OTU number, Simpson's index, and phylogenetic diversity). In other experimental variants, no significant differences are noted.

Table 2. Microbial biodiversity coefficients in the analyzed variants after 168 h of the biodegradation process.

Diversity Index	NP0	NP1	NP2	NP3
OTU number	96 ± 4	101 ± 5	100 ± 2	85 ± 4
Simpson's index	0.85 ± 0.02	0.84 ± 0.02	0.84 ± 0.01	0.79 ± 0.02
Phylogenetic diversity	4.41 ± 0.07	4.57 ± 0.11	4.43 ± 0.08	4.26 ± 0.07

The analysis of the taxonomic structure of bacterial populations shows significant differences between the variants characterized by a low level of nitrogen compounds (NP0 and NP1), and the variants in which the amount of these compounds is at the optimal, and over-optimal, level (NP2 and NP3). In the samples with the levels of nitrogen compounds below the optimal concentration, the domination of the Sphingobacteria (26% for NP0 and 25% for NP1) and Alphaproteobacteria (24% for NP0 and 27% for NP1) is noted. The variants with a lower ratio of carbon to nitrogen show a much higher abundance of the Gammaproteobacteria (39% for NP2 and 41% for NP3) and Betaproteobacteria (24% for NP2 and 28% for NP3). A detailed taxonomic analysis is presented in Figure 4. It should be mentioned that the share of the orders Sphingobacteriales and Xanthomonadales decreases significantly in variants with optimal and excessive levels of nitrogen compounds. The Burkholderiales order shows the opposite trend. Moreover, excessive biostimulation contributes to a large decrease in the share of the Pseudomonadales and Rhizobiales orders, compared to the trials with optimal, or suboptimal, supplementation.

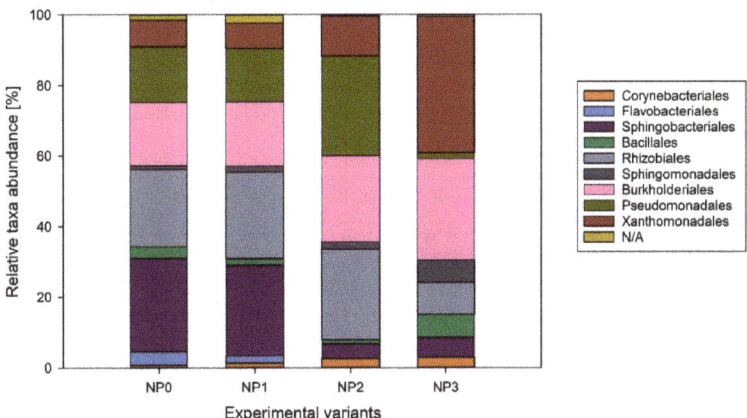

Figure 4. The relative abundance of the bacterial orders in the experimental variants after 168 h of the biodegradation process. Taxa with relative abundance below 1% are excluded from the analysis.

The linear discriminant analysis effect size (LEfSe) allows for the selection of the most differentially abundant taxa between a deficient nitrogen level and a sufficient nitrogen level in experimental variants. (Figure 5). The taxa particularly responsive to changes in the supply of nitrogen compounds include Gammaproteobacteria and Betaproteobacteria, as well as the orders Xanthomonadales, Burkholderiales, Sphingomonadales, Flavobacteriales, and Sphingobacteriales.

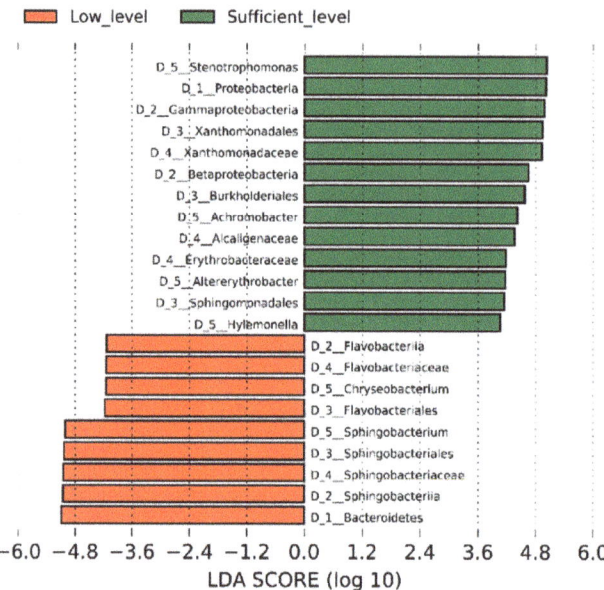

Figure 5. The linear discriminant analysis effect size (LEfSe) analysis of predicted microbial taxa in deficient-nitrogen-level (red) and sufficient-nitrogen-level (green) samples.

3.4. Prediction of Metabolic Properties

The analysis of the predicted metabolic potential for the degradation of PAH compounds in individual bacterial taxa is presented in Figure 6. In all experimental variants, the orders of Sphingobacteriales and Rhizobiales are characterized by a high-predicted representation of genes encoding enzymes involved in PAH biodegradation.

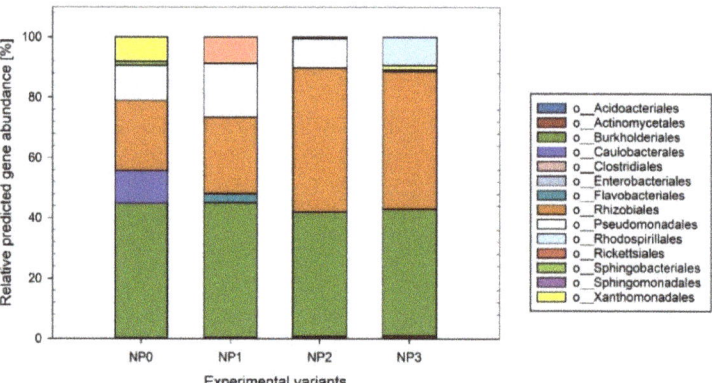

Figure 6. The relative predicted abundance of genes participating in PAH degradation in microbial orders in analyzed experimental variants.

4. Discussion

Biostimulation is considered a promising method of supporting hydrocarbon biodegradation processes, by eliminating element deficiencies, mainly nitrogen and phosphorus [12,14–16]. Biostimulation can be combined with bioaugmentation techniques, which use microorganisms characterized by a high biodegradation potential, in both single strain and consortia. These microorganisms are often isolated from permanently contaminated environments, and they are well described in the literature [12,17,18]. Methods using the enzymatic potential of microbes are considered effective for the degradation of most hydrocarbon fractions. However, there are some limitations in the effective decomposition of aromatic compounds, which belong to the fraction that is extremely resistant to biological degradation [40–42]. Similar conclusions are noted in our research, where the aromatic and polyaromatic fractions are the group of compounds most difficult to decompose. The conducted experiment confirms the increase in the efficiency of biological decomposition of the most problematic fraction, PAH, under the influence of low and optimal supplementation with nitrogen compounds (NP1 and NP2 variants). Interestingly, excessive biostimulation contributes to the inhibition of biological decomposition of all analyzed fractions of hydrocarbon compounds. Similar trends are mentioned in other studies [43,44]. Uncontrolled biostimulation can be counterproductive in the context of hydrocarbon remediation.

The decrease in biodegradation efficiency due to the excessive supply of nitrogen compounds is associated with a decrease in metabolic activity, which is noted both after 24 h (by 3.2%) and after 168 h (by 17.6%) of the experiment. The toxic effect of excessive concentrations of nitrogen compounds on the population of microorganisms could be related to the progressive accumulation of their metabolism products, such as ammonia and ammonium ions, or its oxidation products—nitrates. The phenomenon of the toxicity of ammonia and ammonium ions is very well characterized in the context of the impact on higher organisms, while, in the field of microbiology, this topic still requires detailed research [45]. Moreover, the research by Leejeerajumnean et al. (2000) shows that microorganisms can differ significantly in the level of tolerance to the action of ammonia compounds [46]. This fact may be related to the decrease in alpha-biodiversity observed in the NP3 variant, which suggests the presence of an important selection factor. It is believed that microbial populations have a relatively high metabolic flexibility and functional redundancy; however, they react relatively quickly, with a change in taxonomic structure, to the action of variable, selective environmental factors [47]. This phenomenon is confirmed in the conducted genetic analyses. Changes in the dominant bacterial classes in variants with a moderate and high supply of nitrogen compounds (NP2 and NP3) are observed. There is a dominance of the Gammaproteobacteria and Betaproteobacteria classes, as well as the Xanthomonadales and Burkholderiales orders, which suggests that these classes and orders are particularly important during the population response to changes in the supply of nitrogen compounds. Similar trends are observed in an earlier study [48]. The researchers find an increase in the share of Gammaproteobacteria and Xanthomonadales in soils subjected to a long-term fertilization process with nitrogen compounds. Therefore, it can be assumed that these changes are not accidental.

It can be also postulated that those dominant taxa, which are also the most differentially abundant between deficient and sufficient nitrogen variants, such as Xanthomonadales and Burkholderiales, are more competitive than the other bacteria groups in the use of nitrogen compounds. In an earlier study, Sun et al. (2021) emphasize that long-term exposure to nitrogen compounds results in the dominance of microbial groups associated with ureolysis, and the nitrification and denitrification processes. It is found that the microbial groups that harbor the same or similar functional genes involved in nitrogen use can coexist together. This functional redundancy is well described in the literature, and it is believed that it may increase the stability of microbial metapopulations in the long term [49].

The LEfSe analysis identifies the most significant taxonomic differences between two classes of nitrogen supply (deficient and sufficient). The Xanthomonadales, Burkholderiales, and Sphingomonadales orders may be of key importance in the biological degradation of hydrocarbon fractions in environments with a periodically variable supply of nitrogen compounds, as well as in intensive biostimulation technologies. However, due to their high sensitivity to the variability of environmental parameters, this groups of microorganisms should not play a bioindicative role.

PICRUSt analysis reveals that the Sphingobacteriales and Rhizobiales orders have a high genetic potential for biodegradation of the PAH fraction. The nitrogen-fixing

Rhizobiales order seems to be particularly important. The decrease in its share in the NP3 variant, found after 168 h of biodegradation, can be related to the reduced degradation efficiency of PAH. The biodegradation potential of PAH by microorganisms belonging to the Rhizobiales order is described in the literature. It is believed that Rhizobiales effectively support the process of PAH bioremediation in environments with a deficiency of nitrogen compounds, due to the activity of genes responsible for nitrogen binding [50]. It can be suggested that the quantitative analysis of the Rhizobiales order is helpful in estimating the effectiveness of PAH natural attenuation in areas with a deficiency of nitrogen compounds, and the validity of bioremediation techniques supported by biostimulation. However, it should be mentioned that the process of biological decomposition of polycyclic aromatic hydrocarbons is a complex process. Many studies indicate the importance of cometabolism and metapopulation interactions in the degradation of this group of xenobiotics [51–53]. Therefore, it is not possible to indicate only one group of microorganisms that clearly determines the effectiveness of biological decomposition of PAH.

Based on the current knowledge, it is difficult to make any hypotheses regarding the mechanisms that drive the observed population changes of microorganisms and modifications of the genetic pool in the context of the biological decomposition of hydrocarbon compounds, including PAH. However, it should be emphasized that properly optimized supplementation with nitrogen compounds has a positive effect on the processes of hydrocarbon remediation, if the dose of the biostimulator is optimally selected. Careless application of nitrogenous substances above optimal levels may contribute to the inhibition of biodegradation processes.

5. Conclusions

A properly optimized technology of biostimulation with nitrogen compounds may show a beneficial effect on the biological degradation of one of the most difficult to decompose fractions—PAH. However, excessive supplementation with nutrients may reduce the efficiency of the process, due to decreased metabolic activity of microorganisms and lower biodiversity. Changes in the C:N ratio result in a disturbance of the taxonomic structure of the bacteria population. It also affects the genetic potential of the metapopulation for the PAH degradation. Microorganisms belonging to the Xanthomonadales, Burkholderiales, Sphingomonadales, Flavobacteriales, and Sphingobacteriales orders have a high predicted potential for PAH biodegradation, and show a high quantitative fluctuation under the influence of nitrogen compounds. It can be suggested that the design of a molecular diagnostic test allowing the assessment of the genetic potential of environment by biostimulation, or bioaugmentation supported by biostimulation, should be based on a comprehensive analysis of many groups of microorganisms. Moreover, this analysis should take into account the observed possibility of fluctuations in the shares of the taxa mentioned above.

Author Contributions: Conceptualization, J.S.-P., P.C. and A.P.-C.; methodology, J.S.-P., J.C., W.J. and Ł.W.; validation, J.S.-P., J.C. and W.J.; formal analysis, J.S.-P. and A.P.-C.; investigation, J.S.-P., J.C., W.J., Ł.W., P.C. and A.P.-C.; data curation J.S.-P., J.C., W.J. and Ł.W.; writing—original draft preparation, J.S.-P.; writing—review and editing P.C. and A.P.-C.; visualization, J.S.-P.; supervision, P.C. and A.P.-C.; project administration, A.P.-C. All authors have read and agreed to the published version of the manuscript.

Funding: This research was funded by National Science Centre in Poland in the years 2014–2018, with the research project Opus no. 2013/11/B/NZ9/01908. Publication was financed within the framework of the Polish Ministry of Science and Higher Education's program: "Regional Excellence Initiative" in the years 2019–2022 (No. 005/RID/2018/19), financing amount 1,200,000,000 PLN.

Institutional Review Board Statement: Not applicable.

Informed Consent Statement: Not applicable.

Data Availability Statement: Raw sequence data were deposited in Sequence Read Archive (SRA), as project no PRJNA831882.

Conflicts of Interest: The authors declare no conflict of interest.

References

1. Das, N.; Chandran, P. Microbial degradation of petroleum hydrocarbon contaminants: An overview. *Biotechnol. Res. Int.* **2011**, *2011*, 941810. [CrossRef]
2. Chandra, S.; Sharma, R.; Sharma, A. Application of bioremediation technology in the environment contaminated with petroleum hydrocarbon. *Ann. Microbiol.* **2013**, *63*, 417–431. [CrossRef]
3. Truskewycz, A.; Gundry, T.D.; Khudur, L.S.; Kolobaric, A.; Taha, M.; Aburto-Medina, A.; Ball, A.S.; Shahsavari, E. Petroleum Hydrocarbon Contamination in Terrestrial Ecosystems-Fate and Microbial Responses. *Molecules* **2019**, *24*, 3400. [CrossRef]
4. Abha, S.; Singh, C.S. Hydrocarbon pollution: Effects on living organisms, remediation of contaminated environments, and effects of heavy metals co-contamination on bioremediation. In *Introduction to Enhanced Oil Recovery (EOR) Processes and Bioremediation of Oil-Contaminated Sites*; Romero-Zerón, L., Ed.; InTech Open: Rijeka, Croatia, 2012.
5. Szczepaniak, Z.; Cyplik, P.; Juzwa, W.; Czarny, J.; Staninska, J.; Piotrowska-Cyplik, A. Antibacterial effect of the Trichoderma viride fungi on soil microbiome during PAH's biodegradation. *Int. Biodeterior. Biodegrad.* **2015**, *104*, 170–177. [CrossRef]
6. Szczepaniak, Z.; Czarny, J.; Staninska-Pięta, J.; Lisiecki, P.; Zgoła-Grześkowiak, A.; Cyplik, P.; Chrzanowski, Ł.; Wolko, Ł.; Marecik, R.; Juzwa, W.; et al. Influence of soil contamination with PAH on microbial community dynamics and expression level of genes responsible for biodegradation of PAH and production of rhamnolipids. *Environ. Sci. Pollut. Res.* **2016**, *23*, 23043–23056. [CrossRef]
7. Al-Hawash, A.B.; Dragh, M.A.; Li, S.; Alhujaily, A.; Abbood, H.A.; Zhang, X.; Ma, F. Principles of microbial degradation of petroleum hydrocarbons in the environment. *Egypt. J. Aquat. Res.* **2018**, *44*, 71–76. [CrossRef]
8. Staninska-Pięta, J.; Piotrowska-Cyplik, A.; Juzwa, W.; Zgoła-Grześkowiak, A.; Wolko, Ł.; Sydow, Z.; Kaczorowski, Ł.; Powierska-Czarny, J.; Cyplik, P. The impact of natural and synthetic surfactants on bacterial community during hydrocarbon biodegradation. *Int. Biodeterior. Biodegrad.* **2019**, *142*, 191–199. [CrossRef]
9. Steliga, T. Ocena efektywności biodegradacji węglowodorów ropopochodnych w zastarzałym odpadzie z dołu urobkowego Graby-59 w warunkach przemysłowych metodą in-situ. *Nafta-Gaz* **2014**, *70*, 351–364.
10. Sihag, S.; Pathak, H.; Jaroli, D.P. Factors affecting biodegradation of polyaromatic hydrocarbons. *Int. J. Pure Appl.* **2014**, *2*, 185–202.
11. Ouriache, H.; Moumed, I.; Arrar, J.; Abdelkader, N.; Lounici, H. Influence of C/N/P ratio evolution on biodegradation of petroleum hydrocarbons-contaminated soil. *ALJEST* **2020**, *6*, 1604–1611.
12. Tyagi, M.; da Fonseca, M.R.; de Carvalho, C.C.C.R. Bioaugmentation and biostimulation strategies to improve the effectiveness of bioremediation processes. *Biodegradation* **2011**, *22*, 231–241. [CrossRef]
13. Adams, G.O.; Fufeyin, P.T.; Okoro, S.E.; Ehinomen, I. Bioremediation, biostimulation and bioaugmentation: A review. *Int. J. Environ. Bioremediat. Biodegrad.* **2015**, *3*, 28–39. [CrossRef]
14. Agarry, S.E.; Owabor, C.N. Anaerobic bioremediation of marine sediment artificially contaminated with anthracene and naphthalene. *Environ. Technol.* **2011**, *32*, 1375–1381. [CrossRef]
15. Silva-Castro, G.A.; Rodelas, B.; Perucha, C.; Laguna, J.; González-López, J.; Calvo, C. Bioremediation of diesel-polluted soil using biostimulation as post-treatment after oxidation with Fenton-like reagents: Assays in a pilot plant. *Sci. Total Environ.* **2013**, *15*, 347–355. [CrossRef]
16. Wu, M.; Dick, W.A.; Li, W.; Wang, X.; Yang, Q.; Wang, T.; Xu, L.; Zhang, M.; Chen, L. Bioaugmentation and biostimulation of hydrocarbon degradation and the microbial community in a petroleum-contaminated soil. *Int. Biodeterior. Biodegrad.* **2016**, *107*, 158–164. [CrossRef]
17. Suja, F.; Rahim, F.; Taha, M.R.; Hambali, N.; Razali, M.R.; Khalid, A.; Hamzah, A. Effects of local microbial bioaugmentation and biostimulation on the bioremediation of total petroleum hydrocarbons (TPH) in crude oil contaminated soil based on laboratory and field observations. *Int. Biodeterior. Biodegrad.* **2014**, *90*, 115–122. [CrossRef]
18. Hamoudi-Belarbi, L.; Demdoum, S.; Medjras, S.; Hamoudi, S. Combination of bioaugmentation and biostimulation as an oil-drilling mud contaminated soil bioremediation treatment. *Appl. Ecol. Environ. Res.* **2019**, *17*, 15463–15475. [CrossRef]
19. Vidali, M. Bioremediation. An overview. *Pure Appl. Chem.* **2001**, *73*, 1163–1172. [CrossRef]
20. Kumavath, R.N.; Deverapalli, P. Scientific swift in bioremediation: An overview. In *Bioremediation*; Patil, Y., Ed.; InTech Open: Rijeka, Croatia, 2013.
21. Walworth, J.; Pond, A.; Snape, I.; Rayner, J.; Ferguson, S.; Harvey, P. Nitrogen Requirements for Maximizing Petroleum Bioremediation in a Sub-Antarctic Soil. *Cold Reg. Sci. Technol.* **2007**, *48*, 84–91. [CrossRef]
22. Ruberto, L.; Vazquez, S.C.; Mac Cormack, W.P. Effectiveness of the natural bacterial flora, biostimulation and bioaugmentation on the bioremediation of a hydrocarbon contaminated Antarctic soil. *Int. Biodeterior. Biodegrad.* **2003**, *52*, 115–125. [CrossRef]
23. Liu, P.W.; Chang, T.C.; Whang, L.M.; Kao, C.H.; Pan, P.T.; Cheng, S.S. Bioremediation of petroleum hydrocarbon contaminated soil: Effects of strategies and microbial community shift. *Int. Biodeterior. Biodegrad.* **2011**, *65*, 1119–1127. [CrossRef]
24. Dolinšek, J.; Goldschmidt, F.; Johnson, D.R. Synthetic microbial ecology and the dynamic interplay between microbial genotypes. *FEMS Microbiol. Rev.* **2016**, *40*, 961–979. [CrossRef]
25. Said, S.B.; Or, D. Synthetic microbial ecology: Engineering habitats for modular consortia. *Front. Microbiol.* **2017**, *8*, 1125. [CrossRef]

26. Czarny, J.; Staninska-Pięta, J.; Piotrowska-Cyplik, A.; Juzwa, W.; Wolniewicz, A.; Marecik, R.; Ławniczak, Ł.; Chrzanowski, Ł. Acinetobacter sp. as the key player in diesel oil degrading community exposed to PAHs and heavy metals. *J. Hazard. Mater.* **2020**, *383*, 121168. [CrossRef]
27. Cheng, X.; Zhao, T.; Fu, X.; Hu, Z. Identification of nitrogen compounds in RFCC diesel oil by mass spectrometry. *Fuel Process. Technol.* **2004**, *85*, 1463–1472. [CrossRef]
28. Caporaso, J.G.; Lauber, C.L.; Walters, W.A.; Brg-Lyons, D.; Huntley, J.; Fierer, N.; Owens, S.M.; Betly, J.; Fraser, L.; Bauer, M.; et al. Ultra-high-throughput microbial community analysis on the Illumina HiSeq and MiSeq platforms. *ISME J.* **2012**, *6*, 1621–1624. [CrossRef] [PubMed]
29. Quast, C.; Pruesse, E.; Yilmaz, P.; Gerken, J.; Schweer, T.; Yarza, P.; Peplies, J.; Glöckner, F.O. The SILVA ribosomal RNA gene database project: Improved data processing and web-based tools. *Nucleic Acids Res.* **2013**, *41*, D590–D596. [CrossRef]
30. DeSantis, T.Z.; Hugenholtz, P.; Larsen, N.; Rojas, M.; Brodie, E.L.; Keller, K.; Huber, T.; Dalevi, D.; Hu, P.; Andersen, G.L. Greengenes, a chimera-checked 16S rRNA gene database and workbench compatible with ARB. *Appl. Environ. Microbiol.* **2006**, *72*, 5069–5072. [CrossRef]
31. Segata, N.; Izard, J.; Waldron, L.; Gevers, D.; Miropolsky, L.; Garret, W.S.; Huttenhower, C. Metagenomic biomarker discovery and explanation. *Genome Biol.* **2011**, *12*, R60. [CrossRef]
32. Staley, C.; Gould, T.J.; Wang, P.; Phillips, J.; Cotner, J.B.; Sadowsky, M.J. Core functional traits of bacterial communities in the Upper Mississippi River show limited variation in response to land cover. *Front. Microbiol.* **2014**, *5*, 414. [CrossRef]
33. Czarny, J.; Staninska-Pięta, J.; Piotrowska-Cyplik, A.; Wolko, Ł.; Staninski, K.; Hornik, B.; Cyplik, P. Assessment of soil potential to natural attenuation and autochthonous bioaugmentation using microarray and functional predictions from metagenome profiling. *Ann. Microbiol.* **2019**, *9*, 945–955. [CrossRef]
34. Staninska-Pięta, J.; Czarny, J.; Piotrowska-Cyplik, A.; Juzwa, W.; Wolko, Ł.; Nowak, J.; Cyplik, P. Heavy Metals as a Factor Increasing the Functional Genetic Potential of Bacterial Community for Polycyclic Aromatic Hydrocarbon Biodegradation. *Molecules* **2020**, *25*, 319. [CrossRef]
35. Ahmad, T.; Gupta, G.; Sharma, A.; Kaur, B.; El-Sheikh, M.A.; Alyemeni, M.N. Metagenomic analysis exploring taxonomic and functional diversity of bacterial communities of a Himalayan urban fresh water lake. *PLoS ONE* **2021**, *16*, e0248116. [CrossRef]
36. Hornik, B.; Czarny, J.; Staninska-Pięta, J.; Wolko, Ł.; Cyplik, P.; Piotrowska-Cyplik, A. The Raw Milk Microbiota from Semi-Subsistence Farms Characteristics by NGS Analysis Method. *Molecules* **2021**, *26*, 5029. [CrossRef]
37. Ijoma, G.N.; Nkuna, R.; Mutungwazi, A.; Rashama, C.; Matambo, T.S. Applying PICRUSt and 16S rRNA functional characterisation to predicting co-digestion strategies of various animal manures for biogas production. *Sci. Rep.* **2021**, *11*, 19913. [CrossRef]
38. Li, J.; Huang, B.; Long, J. Effects of different antimony contamination levels on paddy soil bacterial diversity and community structure. *Ecotoxicol. Environ. Saf.* **2021**, *220*, 112339. [CrossRef]
39. Langille, M.G.; Zaneveld, J.; Caporaso, J.G.; McDonald, D.; Knights, D.; Reyes, J.A.; Clemente, J.C.; Burkepile, D.E.; Thurber, R.L.V.; Knight, R.; et al. Predictive functional profiling of microbial communities using 16S rRNA marker gene sequences. *Nat. Biotechnol.* **2013**, *31*, 814–821. [CrossRef]
40. Seo, J.S.; Keum, Y.S.; Li, Q.X. Bacterial degradation of aromatic compounds. *Int. J. Environ. Res.* **2009**, *6*, 278–309. [CrossRef]
41. Staninska, J.; Szczepaniak, Z.; Staninski, K.; Czarny, J.; Piotrowska-Cyplik, A.; Nowak, J.; Marecik, R.; Chrzanowski, Ł.; Cyplik, P. High Voltage Electrochemiluminescence (ECL) as a New Method for Detection of PAH During Screening for PAH-Degrading Microbial Consortia. *Water Air Soil Pollut.* **2015**, *226*, 270. [CrossRef]
42. Czarny, J.; Staninska-Pięta, J.; Powierska-Czarny, J.; Nowak, J.; Wolko, Ł.; Piotrowska-Cyplik, A. Metagenomic analysis of soil bacterial community and level of genes responsible for biodegradation of aromatic hydrocarbons. *Pol. J. Microbiol.* **2017**, *66*, 345–352. [CrossRef]
43. Carmichael, L.M.; Pfaender, F.K. The effect of inorganic and organic supplements on the microbial degradation of phenanthrene and pyrene in soils. *Biodegradation* **1997**, *8*, 1–13. [CrossRef]
44. Chaîneau, C.H.; Rougeux, G.; Yéprémian, C.; Oudot, J. Effects of nutrient concentration on the biodegradation of crude oil and associated microbial populations in the soil. *Soil Biol. Biochem.* **2005**, *37*, 1490–1497. [CrossRef]
45. Müller, T.; Walter, B.; Wirtz, A.; Burkovski, A. Ammonium toxicity in bacteria. *Curr. Microbiol.* **2006**, *52*, 400–406. [CrossRef]
46. Leejeerajumnean, A.; Ames, J.M.; Owens, J.D. Effect of ammonia on the growth of Bacillus species and some other bacteria. *Lett. Appl. Microbiol.* **2000**, *30*, 385–389. [CrossRef]
47. Avila-Jimenez, M.-L.; Burns, G.; He, Z.; Zhou, J.; Hodson, A.; Avila-Jimenez, J.M.-L.; Pearce, D. Functional Associations and Resilience in Microbial Communities. *Microorganisms* **2020**, *8*, 951. [CrossRef]
48. Li, C.; Yan, K.; Tang, L.; Jia, Z.; Li, Y. Change in deep soil microbial communities due to long-term fertilization. *Soil Biol. Biochem.* **2014**, *75*, 264–272. [CrossRef]
49. Sun, R.; Wang, F.; Hu, C.; Liu, B. Metagenomics reveals taxon-specific responses of the nitrogen-cycling microbial community to long-term nitrogen fertilization. *Soil Biol. Biochem.* **2021**, *156*, 108214. [CrossRef]
50. Shin, B.; Bociu, I.; Kolton, M.; Huettel, M.; Kostka, J.E. Succession of microbial populations and nitrogen-fixation associated with the biodegradation of sediment-oil-agglomerates buried in a Florida sandy beach. *Sci. Rep.* **2019**, *9*, 19401. [CrossRef]
51. Ghosal, D.; Ghosh, S.; Dutta, T.K.; Ahn, Y. Current State of Knowledge in Microbial Degradation of Polycyclic Aromatic Hydrocarbons (PAHs): A Review. *Front. Microbiol.* **2016**, *7*, 1369. [CrossRef]

52. Gupta, G.; Kumar, V.; Pal, A.K. Microbial Degradation of High Molecular Weight Polycyclic Aromatic Hydrocarbons with Emphasis on Pyrene. *Polycycl. Aromat. Compd.* **2019**, *39*, 124–138. [CrossRef]
53. Zhang, S.; Hu, Z.; Wang, H. Metagenomic analysis exhibited the co-metabolism of polycyclic aromatic hydrocarbons by bacterial community from estuarine sediment. *Environ. Int.* **2019**, *129*, 308–319. [CrossRef] [PubMed]

Article

Biodegradation of Olive Mill Effluent by White-Rot Fungi

Ana Isabel Díaz, Marta Ibañez, Adriana Laca * and Mario Díaz

Department of Chemical and Environmental Engineering, Faculty of Chemistry, University of Oviedo, C/Julián Clavería, s/n, 33071 Oviedo, Spain; anabel.dglez@gmail.com (A.I.D.); martaibanezlazaro@gmail.com (M.I.); mariodiaz@uniovi.es (M.D.)
* Correspondence: lacaadriana@uniovi.es; Tel.: +34-985-10-29-74

Abstract: The liquid fraction from the two-phase extraction process in the olive industry (alperujo), is a waste that contains lignocellulosic organic matter and phenolic compounds, difficult to treat by conventional biological methods. Lignocellulosic enzymes from white-rot fungi can be an interesting solution to break down these recalcitrant compounds and advance the treatment of that waste. In the present work the ability of *Phanerochaete chrysosporium* to degrade the abovementioned liquid waste (AL) was studied. Experiments were carried out at 26 °C within the optimal pH range 4–6 for 10 days and with and without the addition of glucose, measuring the evolution of COD, BOD_5, biodegradability index, reducing sugars, total phenolic compounds, and colour. The results obtained in this study revealed the interest of *Phanerochaete chrysosporium* for an economical and eco-friendly treatment of alperujo, achieving COD and colour removals around 60%, and 32% of total phenolic compounds degradation, regardless of glucose addition.

Keywords: alperujo; olive mill waste; bioremediation; *Phanerochaete chrysosporium*; fungal treatment

1. Introduction

The Mediterranean region is the main producer of olive oil, concentrating more than 95% of the world's olive trees. Within this area, Spain represents around half of total world manufacturing, considering this industry as one of the most important agri-food sectors for this country [1,2].

Presently, the main industrial process to obtain olive oil is continuous extraction by two-phase or three-phase systems. Depending on the oil extraction system used, considerable amounts of solid and liquid waste as final products are generated. Although the three-phase extraction process mainly generates alpechin and pomace as final residues, the main waste stream from two-phase extraction system is the alperujo (AL) [3]. The pomace is a solid waste composed of the pulp and pits of the olive, commonly employed as fertilizer, biofuel production, or animals feed. The alpechin corresponds with the liquid effluent, composed of water and minerals and characterized by a high organic matter load. This waste stream is considered, together with the AL, a highly polluting residue, for which reuse is not an easy task so its revalorization is still being investigated [4,5]. Due to the global increase of olive oil demand, the excessive amount of waste streams generated throughout the olive oil industry is a growing problem that poses an environmental challenge [6].

In Spain, the two-phase olive oil extraction system is used in approximately 90% of olive mills [4]. The application of this extraction system generates about 800 kg of AL per ton of processed olive, which represents an annual production of around four million tons for the Spanish oil industry [3]. The AL obtained is a semi-solid waste stream composed of vegetable water and olive pomace with high moisture content (60%) that still contains a certain amount of oil [7]. The AL is subjected to a second centrifugation, to obtain a pomace oil. The resulting residue is usually dried in rotary heat dryers at high temperatures and the by-product is subjected to an extraction with hexane to recover more oil, which requires a large amount of energy and incurs high costs [8,9]. Dried or wet AL can also be used in composting processes. However, due to its low porosity, the addition of bulking agents such as bark chips or cotton gin is necessary [9,10]. In addition, its high content of lignocellulosic compounds and polyphenols, which are toxic to animal cells, plants, insects, and microorganisms, is an important drawback [11]. Discharging these wastes without treatment would cause serious damage to aquatic systems, such as the reduction of soluble oxygen. Furthermore, its strong odour would also cause serious problems for the population living near the discharge area [12,13]. Therefore, the removal of the pollutant compounds of the AL, and therefore favouring of the subsequent biological/physical treatment, is one of the main problems that the olive oil industry must confront.

The literature has been mainly focused on the treatment of olive mill wastes coming from the three-phase extraction system, with the aim of removing the organic and phenolic compounds and improving the biodegradability of the effluent. Regarding biological methods, anaerobic digestion or aerobic activated sludge processes have mainly been applied for the treatment of olive mill wastewater (OMW). However, these methods are not usually applied directly to the effluent due to the presence of recalcitrant molecules, polyphenols, the low nutrient load, and the acidic pH of the waste, which make treatment difficult [14]. Previous studies have reported an improvement in the efficiency of these processes reducing the acidity, adding nutrients such as cobalt, or extracting polyphenols before biodegradation [15]. Due to the high level of antimicrobial compounds present in OMW, the acclimatization of biomass or the use of physical $CaCO_3$ supports has also been required to improve the biodegradation and methanization process [16]. Therefore, the traditional biological methods are not as effective as would be desirable. Physical–chemical methods such as nanofiltration, ultrafiltration, ultrasound, hydrothermal carbonization, and different advances oxidation processes have been reported for the treatment of OMW reducing its chemical oxygen demand (COD) and phenol content [17,18]. However, these techniques have several drawbacks, such as the addition of chemicals, fouling of the membrane, and high pressure and temperature conditions [5]. Therefore, the search for alternative methods that allow the treatment of OMW and AL in an economic and eco-friendly way is crucial.

The use of fungi has been described as a promising alternative over the use of bacteria for OMW treatment due to its ability to grow under adverse conditions and to produce a great variety of extracellular enzymes that make possible the degradation of recalcitrant compounds [19]. In this way, fungi can break down the complex recalcitrant compounds making them more assimilable to be used by themselves or by the bacteria in a subsequent treatment [20,21]. The OMW treatment by fungi has mainly focused of removing COD, phenolic content, and colour, as well as obtaining by-products with biotechnological interest, such as fungal enzymes [22,23]. White-rot fungi can degrade the lignin present in lignocellulosic wastes due to the release of enzymes, mainly lignin peroxidase and manganese peroxidase [24]. These fungi have been investigated for the treatment of recalcitrant compounds and colour degradation of OMW, obtaining good results. For example, Ntougias et al. [25], who studied the capacity of several strains of *Pleurotus* and *Ganoderma* fungi to treat OMW, reported significant removals of COD, TOC, and phenolic compounds, as well as a reduction of the toxicity of the effluent. COD degradations around 50% have been achieved by the fungus *Phanerochaete chrysosporium* immobilized on loofah [26]. Great removals of colour, phenolic compounds and COD have also been obtained when OMW was treated with fungi from genus *Aspergillus* [22,27].

As far as we know, the bioremediation with fungi has been mainly applied to treat OMW or pomace coming from the three-extraction system. However, its application to AL waste, obtained from the two-phase extraction system, has been hardly studied. Therefore, the main objective of this study was to investigate the capability of the white-rot fungus *Phanerochaete chrysosporium* to treat AL waste, to reduce its COD, colour, and phenolic compounds.

2. Materials and Methods

2.1. Sample Description

The AL used for this work corresponds with the semi-solid effluent generated during olive oil extraction by a two-phase extraction system. The sample was collected from an olive oil factory sited in Sevilla, Spain. For the fungal treatment, the sample was mixed with distilled water in a ratio 1:20 (p/v). The mix was filtered using a 1.5 mm mesh sieve to remove the rest of the peel and pit of the olives. After that, the effluent was centrifuged for 10 min at $9000\times g$ and the supernatant was filtered by a cellulose filter (10–20 μm). This diluted AL was used for the subsequent fungal treatments. The characteristics of diluted AL are shown in Table 1.

Table 1. Characteristics of the diluted AL.

Parameter	Value
pH	4.6 ± 0.01
sCOD (mg O_2/L)	4854 ± 19
sBOD (mg O_2/L)	408 ± 14
Biodegradability Index (B.I.)	0.080 ± 0.003
Reducing sugars (mg/L)	578 ± 24
Total phenolic compounds (mg/L)	134 ± 4
Colour index (C.I.)	1.60 ± 0.04
Total Suspended Solids (mg/L)	2475 ± 21
Fixed Suspended Solids (mg/L)	375 ± 12
Volatile Suspended Solids (mg/L)	2100 ± 28

2.2. Fungal Pellet Obtention

The white-rot fungus, *Phanerochaete chrysosporium* Burdsall 1974 was used. The freeze-dried strain (CECT 2798 from Spanish Type Culture Collection) was recovered in aseptic conditions by adding 100 μL of the resuspended fungus to 10 mL of malt extract (ME). Then, a Petri plate of 1.5% malt extract agar (MEA) was inoculated with 100 μL of this suspension and incubated at 26 °C for 6 days. Two subcultures of the fungus were necessary before use in the biological treatment. Fungal subcultures were routinely made every month to conserve the strain.

The methodology described by Díaz et al. [28] was followed to obtain the fungus pellets. To this aim, five cylinders of 1 cm diameter from the growing zone of inoculated plates were used to inoculate 500 mL Erlenmeyer flasks containing 150 mL of sterilised malt extract broth (VWR Chemicals BDH), with a pH between 4.5 and 5. The inoculated flasks were incubated at 26 °C and 135 rpm for 6 days. The fungal mycelial obtained after this process was separated with a sieve and homogenized with 0.8% NaCl (w/v) in a ratio of 1:3 (w/v). An amount of 600 μL of resulting suspension was used to inoculate a 1 L Erlenmeyer flask with 250 mL of sterilised ME. Finally, the inoculated ME was incubated at 26 °C and 135 rpm for 6 days. After that time, pellets were obtained, removed with a sieve, and preserved in 0.8% NaCl (w/v) solution at 4 °C until use.

2.3. Fungal Treatment

Several batch tests were carried out to treat diluted AL with *P. chrysosporium*. All the experiments were performed using 1 L Erlenmeyer flask with 250 mL of AL effluent.

- Test E1 and E2 were inoculated with the fungus pellet (3 g/L of dry matter), with the only difference that E2 was supplied with 3 g/L of glucose.
- Test C1 and C2 were used as control without fungus inoculation, without and with glucose addition, respectively.

The flasks were incubated at 26 °C and in an orbital shaking (150 rpm) for 10 days. During the treatment, the pH values were maintained within the range 5–7 by adding NaOH 0.5 M or HCl 0.5 M to ensure the optimal range for the fungus enzymatic system. Samples taken periodically were centrifuged at $15,000\times g$ for 15 min and the supernatant were conserved at 4 °C until analysed. The experiments were carried out in duplicate. Data shown in Results and Discussion section are the average values of both experiments. In all cases, standard deviations were lower than 15% with respect to average value.

2.4. Analytical Methods

2.4.1. Determination of sCOD, sBOD$_5$, and Biodegradability Index

The concentration of soluble COD (sCOD) was spectrophotometrically measured (at 600 nm) by dichromate method according to Standard Methods [29], using a DR2500 spectrophotometer (Hach Company). Soluble biochemical oxygen demand (sBOD$_5$) was determined using a manometric respirometry measurement system (Lovibond Water Testing BD 600) and biodegradability index (BI) was calculated as the ratio of sBOD$_5$ over sCOD.

2.4.2. Determination of Colour and pH

The change in the colour of the AL was determined by means of the colour index (CI), which is defined according to Equation (1) [30].

$$CI = \frac{SAC_{436}^2 + SAC_{525}^2 + SAC_{620}^2}{SAC_{436} + SAC_{525} + SAC_{620}} \quad (1)$$

Spectral absorbance coefficients (SAC) are defined as the ratio of the values of the respective absorbance over the cell thickness. The absorbances were measured at 436, 525 and 620 nm using a UV/vis spectrophotometer (Thermo Scientific, Heλios γ). The value of pH was measured by means of a pH-meter (Basic-20 Dilabo).

2.4.3. Determination of Total Reducing Sugars

The total reducing sugars concentration was determined by the dinitrosalicylic acid (DNS) method with glucose as standard, according to the Miller's method [31]. The absorbance of samples was measured at 540 nm. The glucose was used as standard.

2.4.4. Determination of Total Phenolic Compounds

The total phenolic compounds were determined by the Folin–Ciocalteu method in dark conditions, according to Moussi et al. [32], using gallic acid as standard. In this procedure, 400 µL of sample were mixed with 3 mL of Folin–Ciocalteu reagent (previously diluted 1:10 with distilled water). This mixture was maintained at 22 °C for 5 min. After that, 3 mL of sodium bicarbonate (NaHCO$_3$ 6 g/100 mL) were added, and the sample was again incubated at 22 °C for 90 min. After incubation, the absorbance was measured at 725 nm.

2.4.5. Determination of Moisture, TSS, FSS, and VSS

Total suspended solids (TSS), fixed suspended solids (FSS) and moisture were measured according to Standard Methods [29]. The volatile suspended solids were calculated as the difference between TSS and FSS.

3. Results and Discussion

3.1. Removal of Organic Matter

The evolution of sCOD concentration during the fungal treatments is shown in Figure 1.

The initial sCOD value in the AL effluent was 4854 mg/L, which increased to 9243 mg/L after glucose addition. For C1 and C2 test, which were carried out without fungal inoculation, minor sCOD removals were observed. This degradation was carried out by the endogenous microbiota present in the effluent. With respect to test C1, no change was observed during the first two days and a 27% elimination of sCOD was achieved after four days of incubation. Afterwards, the sCOD value remained almost constant, reaching a final sCOD degradation of 30%. In the case of C2, where glucose where added, sCOD degradation did not occur until the 4th day. The final percentage of sCOD removal was similar to that achieved with C1. However, it is necessary to point out that in C2 the final sCOD concentration was higher than the sCOD of AL before being supplemented with glucose. Therefore, the endogenous microorganisms were not able to assimilate even the sCOD provided by glucose added.

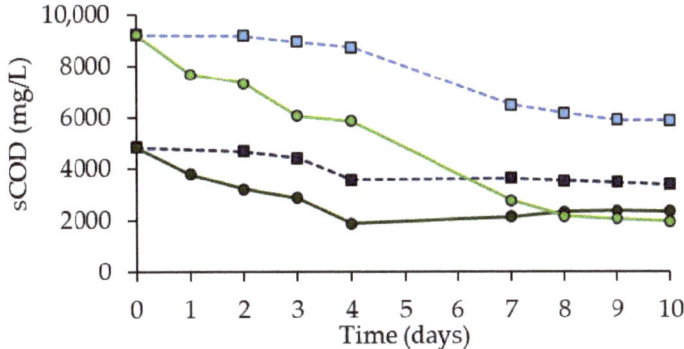

Figure 1. Changes in sCOD concentration during biological treatment. The dashed lines shown the non-inoculated tests C1 (■) and C2 (■), used as controls. The solid lines shown the inoculated tests E1 (●) and E2 (●). The standard deviation (SD) of the experimental data were in all cases less than 6.5% of mean value.

The addition of *P. chrysosporium* in E1 and E2, caused a fast decrease in sCOD from the beginning of the treatment, obtaining sCOD removals of 51% and 59%, respectively, after 10 days of treatment. In the experiment E1, carried out without the addition of glucose, 61% of the initial sCOD was degraded in only 4 days of treatment, which duplicate the degradation efficiency reached in C1 by endogenous microorganisms. Thus, the fungal inoculation gave an average rate of sCOD degradation of 0.51 mg/(L min) during the first 4 days, whereas the average rate in C1 was only 0.22 mg/(L min). Regarding test E2, again higher sCOD removals were obtained compared with the non-inoculated test C2. As in E1, the inoculation of the fungus duplicated the sCOD removal rate, which increased up to 0.61 mg/(L min) during first 8 days, whereas in C2 it was only of 0.27 mg/(L min).

To estimate the fungus growth, the TSS at the beginning of the experiments and after 10 days were measured. Data are shown in Table 2.

Table 2. TSS for control (C1 and C2) and inoculated (E1 and E2) tests at initial and final times of the fungal treatment.

Sample	Initial TSS (g/L)	Final TSS (g/L)	Increase (g/L)
C1	2.47 ± 0.01	3.06 ± 0.01	0.61 ± 0.01
E1	6.87 ± 0.01	8.21 ± 0.01	1.34 ± 0.01
C2	2.47 ± 0.01	5.16 ± 0.01	2.71 ± 0.01
E2	6.87 ± 0.02	10.16 ± 0.02	3.29 ± 0.02

As can be seen, in the control tests (C1 and C2), the supplementation with glucose increased the growth of the endogenous microbiota. Moreover, in the inoculated tests (E1 and E2), the TSS increases were higher than in the controls, which can be explained by the fungus growth. Comparing the increase in TSS observed in controls and inoculated tests, it can be estimated that fungus growth was similar in E1 and E2, around 0.6–0.7 g/L (dry matter), which is in agreement with the fact that final sCOD removals were also similar. Therefore, in this case, the addition of glucose was not effective for the AL treatment.

Results for sCOD removals here obtained were higher than have been previously reported. Aloui et al. [33], reported that a 44% of COD removal was achieved by a solid-state fermentation of AL using *P. chrysosporium* in a support of sugarcane bagasse. Ahmadi et al. [26] achieved a COD degradation around 50% using this fungus immobilized on loofah. Nogueira et al. [34] reported COD removals efficiencies lower than 44% for *P. chrysosporium* for an OMW pre-treat by photocatalytic oxidation.

The initial concentration of sBOD$_5$ was 408 mg/L, with a biodegradability index of 0.08 (See Figure 2), which means that AL effluent has very low biodegradability. In the experiments C2 and E2, the initial biodegradability index (BI) was higher (0.13), as a consequence of the glucose addition.

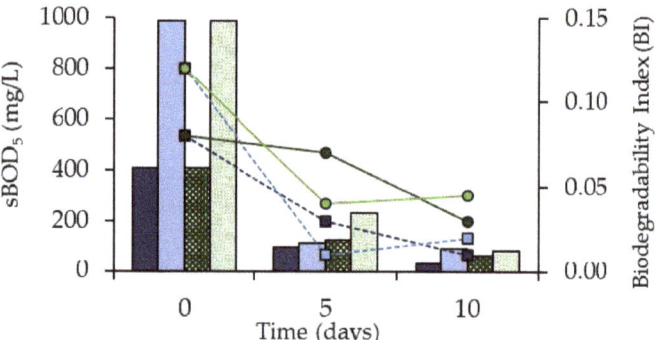

Figure 2. Changes in sBOD$_5$ concentration and biodegradability index for the different experiments at initial, intermediate, and final time of the treatment. Bars corresponds with sBOD$_5$ concentration for non-inoculated tests C1 (■) and C2 (■), and inoculated tests E1 (■) and E2 (■). The dashed lines shown the biodegradability index (BI) for non-inoculated tests C1 (■) and C2 (■), used as controls, whereas the solid lines represent the inoculated tests E1 (•) and E2 (•).

In all cases, the sBOD$_5$ concentration decreased throughout the fungal treatment, with final values lower than 100 mg/L. Moreover, the BI decreases with the treatment since biodegradable matter was consumed. The *P. chrysosporium* inoculated in E1 and E2 released enzymes able to break down recalcitrant organic matter into compounds more biodegradable. However, the fungus, as well as the endogenous microorganisms, consumed these compounds as they were produced, reducing the sCOD, the sBOD$_5$, and the BI. Regardless, the addition of the fungus gave final BI higher than in the controls, even though it was low. If the enhancing of biodegradability were the objective, for example, as the previous step for the biomethanization process, an alternative could be to directly use the enzymes produced by the fungus instead of inoculating the fungus strain. In this way, the recalcitrant compounds present in the AL effluent could be broken down without the fungus using this organic matter as a nutrient source [35,36]. A sterilisation process may also be necessary to inactivate the endogenous microflora.

The evolution of reducing carbohydrates has been also measured, and results are shown in Figure 3. The reducing sugar concentration of the initial sample was 563 mg/L, and the ratio sBOD$_5$/reducing sugars was 0.7, indicating that a great part of the sBOD$_5$ measured is due to the reducing sugars. As expected, in the samples supplemented with 3 g/L of glucose, the initial concentration increased until 3662 mg/L. For the non-inoculated sample C1, the amount of reducing sugars remained practically stable during the treatment. In contrast, the inoculated samples (E1 and E2) showed a significant decrease in the reducing sugars concentration, with final values of 176 and 140 mg/L, respectively. The initial reducing sugars concentration dropped abruptly in the experiments supplemented with glucose, especially in the one that had been inoculated with the fungus. In this sense, all the glucose that was practically added to test E2 was consumed during the first 24 h, whereas in the supplemented control (C2), the amount of reducing sugars dropped from

3663 mg/L to 1772 mg/L in 24 h, and afterwards remained almost constant, indicating that the endogenous microbiota was not able to degrade all the glucose added. Although the enzymatic activities of the fungus were not measured in this study, the literature has widely reported that the addition of glucose favours the synthesis of fungal enzymes, which in turn are related to the elimination of colour, COD and recalcitrant compounds [37–40]. This fact was reflected in E2, which showed a rapid degradation of reducing sugars, whereas sCOD removal was slower. Probably the fungus decomposed recalcitrant compounds that increased sCOD and, simultaneously, consumed them.

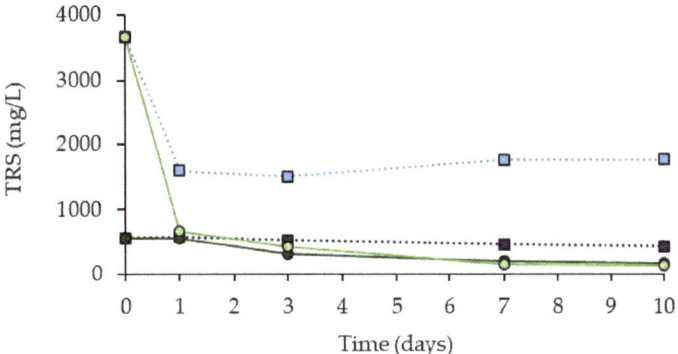

Figure 3. Changes in reducing sugars concentration during biological treatment. The dashed lines shown the non-inoculated tests C1 (■) and C2 (■), used as controls. The solid lines shown the inoculated tests E1 (●) and E2 (●). The standard deviation (SD) of the experimental data were in all cases less than 11% of mean value.

3.2. Removal of Phenolic Compounds

The concentration of total phenolic compounds in the initial AL effluent and in AL effluent after fungal treatment was analysed. The phenolic compounds cause severe pollution of surface and ground water, soils, and vegetation. Its presence has a negative effect on microorganisms due to its high antibacterial activity [35,36].

As is shown in Figure 4, the best efficiencies of phenolic compound removal were reached with the inoculation of *P. chrysosporium* in E1, where around 30% of phenolic compounds were degraded after 10 days of treatment reaching values of 91 mg/L. This percentage of removal was slightly lower when glucose was added (E2), obtaining final removals of 25%. For the non-inoculated samples with fungus (C1 and C2), the amount of phenolic compounds removed was lower, with removal percentage of 12% in both cases. Results proved that the fungus inoculation increases the degradation of phenolic compounds with removal percentages almost three times greater than in the non-inoculated samples. However, higher efficiencies have been reported by other authors when the AL effluent was previously sterilised.

Elisashvili et al. [23], who treated a diluted and sterilised olive pomace effluent by submerged fermentation with *Cerrena unicolor*, reported a removal of phenolic content of around 80%. Additionally, this fungus showed a good capacity to release laccases, which are involved in the degradation of phenolic compounds. The low removals achieved in this study it could be because a non-sterilised AL effluent. Moreover, low laccase activity has been reported for *P. chrysosporium* [24]. García et al. [35] reported a 92% total phenol degradation using *P. chrysosporium* to treat a sterilised OMW supplied with a nitrogen source. Additionally, great phenolic removals were obtained when AL was dried, and the concentrate was treated. For example, Sampedro et al. [37] reported removals around 85% using the fungus *Phlebia* sp. immobilized in polyurethane sponge, while 43% was achieved when the effluent was treated by free mycelia.

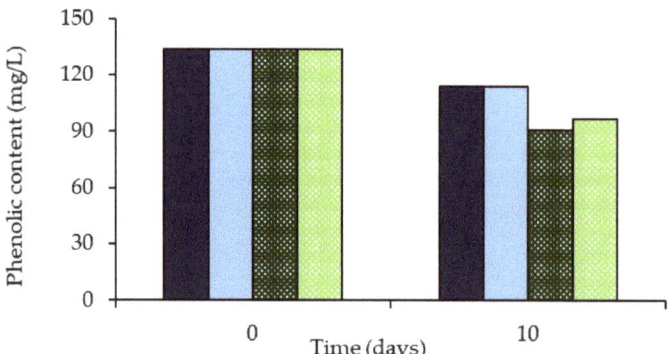

Figure 4. Changes in total phenolic content for non-inoculated tests C1 (■) and C2 (■), and inoculated tests E1 (■) and E2 (■). The standard deviation (SD) of the experimental data were in all cases less than 4% of mean value.

3.3. Removal of Colour

AL waste has a dark brown colour, so its decolourization is important to avoid negative environmental and visual effects. Highly coloured wastewater reduces the passage of light through the water, causing a reduction in photosynthetic activity and, therefore, altering the flora and fauna of the water [38]. The colour index profile is shown in Figure 5. Greater removals were obtained when the fungus was added.

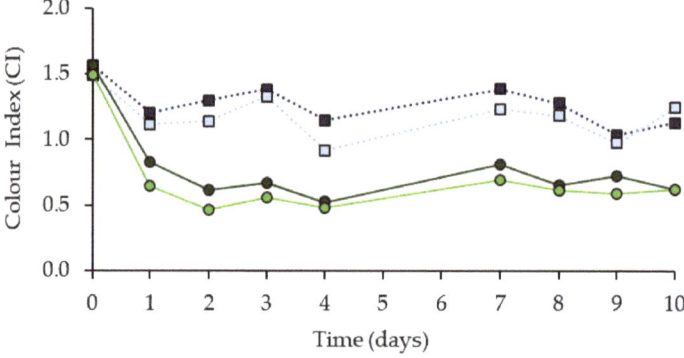

Figure 5. Changes in colour index during biological treatment. The dashed lines shown the non-inoculated tests C1 (■) and C2 (■), used as controls. The solid lines shown the inoculated tests E1 (●) and E2 (●). The standard deviation (SD) of the experimental data were in all cases less than 11.5% of mean value.

In the non-inoculate experiments (C1 and C2), the colour index was also reduced by endogenous microflora, especially during the first 24 h. Afterwards, the C1 slightly changed up and down, reaching final percentages of removal around 20%. When the fungus was inoculated (E1 and E2) the colour index decreased more abruptly especially during the first 24 h. Removals around 70% were obtained in both cases after 4 days. Then, the colour removal slightly increased and remained approximately stable, finally reaching a 60% reduction. The obtained results were in accordance with those found in the literature regarding the colour removals in recalcitrant wastewaters using white-rot fungi. Pakshirajan and Kheria [39] reported colour degradations of 64% after continuous fungal treatment with *P. chrysosporium* of industrial textile wastewaters. Ntougias et al. [25] reported colour removals around 60–65% in OMW using basidiomycetes fungus *Pleurotus* spp. Similar reductions were reported for olive mill effluent treated by an adapted strain of *Trametes versicolor* [40].

Taking into account data reported by the literature and results obtained in this work, the use of white-rot fungi to treat AL could be considered to be a promising treatment technology. Although the removal efficiencies of sCOD, colour and phenolic compounds after treatment with fungi were slightly lower than those reported for OMW treated with physical–chemical treatments, it should be considered that these processes usually present serious drawbacks such as their high cost, bad odour, addition of chemicals, or fouling of the membrane [4,18]. Additionally, chemical oxidation treatments can produce more recalcitrant or toxic intermediate compounds, reducing the effectiveness of the treatment. In contrast, the use of white-rot fungus allows the degradation of a wide range of recalcitrant contaminants due to its ability to release extracellular enzymes, as well as lead to the detoxification of wastewater [25].

4. Conclusions

Biotreatment with white-rot fungus *Phanerochaete chrysosporium* is useful to degrade AL waste. When the non-inoculate AL was incubated at 26 °C, 27% of the sCOD was removed in 4 days, whereas the inoculation of fungus allowed the achievement of a sCOD degradation of 60% over the same time. The addition of glucose as an easy carbon source did not enhance the sCOD degradation. The addition of *P. chrysosporium* also allowed a reduction of the colour index of the residue close to 60%. In all the conditions tested, including a control test without inoculation, a reduction in the sBOD$_5$ and in the reducing sugar content was observed after the treatment. However, the biodegradability index decreased, more highly in the inoculated effluent than those in the absence of inoculation. Finally, the treatment of the diluted AL with the fungus allowed degradation of 32% of the total phenols initially present in the effluent, whereas the endogenous microflora could only degrade around 15% of phenolic content. Results obtained in this study open the possibility of using *P. chrysosporium* fungus in the bioremediation of low-biodegradable wastes from the olive oil industry.

Author Contributions: Conceptualization: A.I.D., M.I., A.L. and M.D.; methodology: A.I.D., M.I., A.L. and M.D.; validation: A.I.D., M.I. and A.L.; formal analysis: A.I.D., M.I. and A.L.; investigation: A.I.D. and M.I.; data curation: A.I.D. and M.I.; writing—original draft preparation: A.I.D.; writing—review and editing: A.L. and M.D.; visualization: A.I.D. and M.I.; supervision: A.L. and M.D.; project administration, M.D.; funding acquisition, M.D. All authors have read and agreed to the published version of the manuscript.

Funding: This work was co-financed by Spanish MINECO (Project CTM2015-63864-R) and Foundation for the Promotion of Applied Scientific Research and Technology in Asturias (Project FC-GRUPIN-IDI/2018/000127). Ana Isabel Díaz acknowledges an FPI grant from Spanish MICINN (BES-2016-077963).

Institutional Review Board Statement: Not applicable.

Informed Consent Statement: Not applicable.

Data Availability Statement: Not applicable.

Conflicts of Interest: The authors declare no conflict of interest.

References

1. Messineo, A.; Maniscalco, M.P.; Volpe, R. Biomethane recovery from olive mill residues through anaerobic digestion: A review of the state of the art technology. *Sci. Total Environ.* **2020**, *703*, 135508. [CrossRef]
2. International Olive Council, Economic Affairs & Promotion Unit. EU Olive Oil Figures. 2020. Available online: https://www.internationaloliveoil.org/wp-content/uploads/2020/12/OT-CE-901-23-11-2020-P.pdf (accessed on 23 March 2021).
3. Alburquerque, J. Agrochemical characterisation of "alperujo", a solid by-product of the two-phase centrifugation method for olive oil extraction. *Bioresour. Technol.* **2004**, *91*, 195–200. [CrossRef]
4. Dermeche, S.; Nadour, M.; Larroche, C.; Moulti-Mati, F.; Michaud, P. Olive mill wastes: Biochemical characterizations and valorization strategies. *Process Biochem.* **2013**, *48*, 1532–1552. [CrossRef]

5. Roig, A.; Cayuela, M.L.; Sánchez-Monedero, M.A. An overview on olive mill wastes and their valorisation methods. *Waste Manag.* **2006**, *26*, 960–969. [CrossRef]
6. Benavente, V.; Fullana, A. Torrefaction of olive mill waste. *Biomass Bioenergy* **2015**, *73*, 186–194. [CrossRef]
7. Sampedro, I.; Aranda, E.; Rodríguez-Gutierrez, G.; Lama-Muñoz, A.; Ocampo, J.A.; Fernández-Bola, J.; García-Romera, I. Effect of a New Thermal Treatment in Combination with Saprobic Fungal Incubation on the Phytotoxicity Level of Alperujo. *J. Agric. Food Chem.* **2011**, *59*, 3239–3245. [CrossRef]
8. Göğüş, F.; Maskan, M. Air drying characteristics of solid waste (pomace) of olive oil processing. *J. Food Eng.* **2006**, *72*, 378–382. [CrossRef]
9. Galliou, F.; Markakis, N.; Fountoulakis, M.S.; Nikolaidis, N.; Manios, T. Production of organic fertilizer from olive mill wastewater by combining solar greenhouse drying and composting. *Waste Manag.* **2018**, *75*, 305–311. [CrossRef]
10. Alburquerque, J.A.; Gonzálvez, J.; García, D.; Cegarra, J. Effects of a compost made from the solid by-product ("alperujo") of the two-phase centrifugation system for olive oil extraction and cotton gin waste on growth and nutrient content of ryegrass (Lolium perenne L.). *Bioresour. Technol.* **2007**, *98*, 940–945. [CrossRef]
11. Justino, C.I.; Duarte, K.; Loureiro, F.; Pereira, R.; Antunes, S.C.; Marques, S.M.; Gonçalves, F.; Rocha-Santos, T.A.P.; Freitas, A.C. Toxicity and organic content characterization of olive mill wastewater undergoing a sequential treatment with fungi and photo-Fenton oxidation. *J. Hazard. Mater.* **2009**, *172*, 1560–1572. [CrossRef]
12. Lee, Z.S.; Chin, S.Y.; Lim, J.W.; Witoon, T.; Cheng, C.K. Treatment technologies of palm oil mill effluent (POME) and olive mill wastewater (OMW): A brief review. *Environ. Technol. Innov.* **2019**, *15*, 100377. [CrossRef]
13. Smeti, E.; Kalogianni, E.; Karaouzas, I.; Laschou, S.; Tornés, E.; de Castro-Català, N.; Anastasopoulou, E.; Koutsodimou, M.; Andriopoulou, A.; Vardakas, L.; et al. Effects of olive mill wastewater discharge on benthic biota in Mediterranean streams. *Environ. Pollut.* **2019**, *254*, 113057. [CrossRef]
14. Gonçalves, C.; Lopes, M.; Ferreira, J.P.; Belo, I. Biological treatment of olive mill wastewater by non-conventional yeasts. *Bioresour. Technol.* **2009**, *100*, 3759–3763. [CrossRef]
15. Pinto-Ibieta, F.; Serrano, A.; Jeison, D.; Borja, R.; Fermoso, F.G. Effect of cobalt supplementation and fractionation on the biological response in the biomethanization of Olive Mill Solid Waste. *Bioresour. Technol.* **2016**, *211*, 58–64. [CrossRef] [PubMed]
16. Hanafi, F.; Belaoufi, A.; Mountadar, M.; Assobhei, O. Augmentation of biodegradability of olive mill wastewater by electrochemical pre-treatment: Effect on phytotoxicity and operating cost. *J. Hazard. Mater.* **2011**, *190*, 94–99. [CrossRef]
17. Kuscu, O.S.; Eke, E. Oxidation of olive mill wastewater by a pulsed high-voltage discharge using oxygen or air. *J. Environ. Chem. Eng.* **2021**, *9*, 104701. [CrossRef]
18. Domingues, E.; Fernandes, E.; Gomes, J.; Castro-Silva, S.; Martins, R.C. Olive oil extraction industry wastewater treatment by coagulation and Fenton's process. *J. Water Process Eng.* **2021**, *39*, 101818. [CrossRef]
19. Rodríguez-Couto, S. Industrial and environmental applications of white-rot fungi. *Mycosphere* **2017**, *8*, 456–466. [CrossRef]
20. Collado, S.; Oulego, P.; Suárez-Iglesias, O.; Díaz, M. Biodegradation of dissolved humic substances by fungi. *Appl. Microbiol. Biotechnol.* **2018**, *102*, 3497–3511. [CrossRef]
21. Rouches, E.; Herpoël-Gimbert, I.; Steyer, J.P.; Carrere, H. Improvement of anaerobic degradation by white-rot fungi pretreatment of lignocellulosic biomass: A review. *Renew. Sustain. Energy Rev.* **2016**, *59*, 179–198. [CrossRef]
22. Salgado, J.M.; Abrunhosa, L.; Venâncio, A.; Domínguez, J.M.; Belo, I. Combined bioremediation and enzyme production by *Aspergillus* sp. in olive mill and winery wastewaters. *Int. Biodeterior. Biodegrad.* **2016**, *110*, 16–23. [CrossRef]
23. Elisashvili, V.; Kachlishvili, E.; Asatiani, M.D. Efficient production of lignin-modifying enzymes and phenolics removal in submerged fermentation of olive mill by-products by white-rot basidiomycetes. *Int. Biodeterior. Biodegrad.* **2018**, *134*, 39–47. [CrossRef]
24. Coconi-Linares, N.; Ortiz-Vázquez, E.; Fernández, F.; Loske, A.M.; Gómez-Lim, M.A. Recombinant expression of four oxidoreductases in *Phanerochaete chrysosporium* improves degradation of phenolic and non-phenolic substrates. *J. Biotechnol.* **2015**, *209*, 76–84. [CrossRef] [PubMed]
25. Ntougias, S.; Baldrian, P.; Ehaliotis, C.; Nerud, F.; Antoniou, T.; Merhautová, V.; Zervakis, G.I. Biodegradation and detoxification of olive mill wastewater by selected strains of the mushroom genera *Ganoderma* and *Pleurotus*. *Chemosphere* **2012**, *88*, 620–626. [CrossRef]
26. Ahmadi, M.; Vahabzadeh, F.; Bonakdarpour, B.; Mehranian, M.; Mofarrah, E. Phenolic removal in olive oil mill wastewater using loofah-immobilized *Phanerochaete chrysosporium*. *World J. Microbiol. Biotechnol.* **2006**, *22*, 119–127. [CrossRef]
27. Abrunhosa, L.; Oliveira, F.; Dantas, D.; Gonçalves, C.; Belo, I. Lipase production by *Aspergillus ibericus* using olive mill wastewater. *Bioprocess Biosyst. Eng.* **2013**, *36*, 285–291. [CrossRef] [PubMed]
28. Díaz, A.I.; Laca, A.; Díaz, M. Fungal treatment of an effluent from sewage sludge digestion to remove recalcitrant organic matter. *Biochem. Eng. J.* **2021**, *172*, 108056. [CrossRef]
29. Baird, R.B.; Rice, C.E.W.; Eaton, A.D. *Standard Methods for Examination of Water and Wastewater*, 23th ed.; American Public Health Association: Washington, DC, USA, 2005.
30. Díaz, A.I.; Oulego, P.; González, J.M.; Laca, A.; Díaz, M. Physico-chemical pre-treatments of anaerobic digestion liquor for aerobic treatment. *J. Environ. Manag.* **2020**, *274*, 111189. [CrossRef] [PubMed]
31. Díaz, A.I.; Laca, A.; Laca, A.; Díaz, M. Treatment of supermarket vegetable wastes to be used as alternative substrates in bioprocesses. *Waste Manag.* **2017**, *67*, 59–66. [CrossRef]

32. Moussi, K.; Nayak, B.; Perkins, L.B.; Dahmoune, F.; Madani, K.; Chibane, M. HPLC-DAD profile of phenolic compounds and antioxidant activity of leaves extract of *Rhamnus alaternus* L. *Ind. Crops Prod.* **2015**, *74*, 858–866. [CrossRef]
33. Aloui, F.; Abid, N.; Roussos, S.; Sayadi, S. Decolorization of semisolid olive residues of "alperujo" during the solid state fermentation by *Phanerochaete chrysosporium*, *Trametes versicolor*, *Pycnoporus cinnabarinus* and *Aspergillus niger*. *Biochem. Eng. J.* **2007**, *35*, 120–125. [CrossRef]
34. Nogueira, V.; Lopes, I.; Freitas, A.C.; Rocha-Santos, T.A.P.; Gonçalves, F.; Duarte, A.C.; Pereira, R. Biological treatment with fungi of olive mill wastewater pre-treated by photocatalytic oxidation with nanomaterials. *Ecotoxicol. Environ. Saf.* **2015**, *115*, 234–242. [CrossRef]
35. García, I.G.; Peña, P.R.R.J.; Venceslada, J.L.L.B.; Martín, A.M.; Santos, M.A.M.; Gómez, E.R. Removal of phenol compounds from olive mill wastewater using *Phanerochaete chrysosporium*, *Aspergillus niger*, *Aspergillus terreus* and *Geotrichum candidum*. *Process Biochem.* **2000**, *35*, 751–758. [CrossRef]
36. Martínková, L.; Kotik, M.; Marková, E.; Homolka, L.; Markov, E. Biodegradation of phenolic compounds by Basidiomycota and its phenol oxidases: A review. *Chemosphere* **2016**, *149*, 373–382. [CrossRef]
37. Sampedro, I.; Cajthaml, T.; Marinari, S.; Stazi, S.R.; Grego, S.; Petruccioli, M.; Federici, F.; D'Annibale, A. Immobilized Inocula of White-Rot Fungi Accelerate both Detoxification and Organic Matter Transformation in Two-Phase Dry Olive-Mill Residue. *J. Agric. Food Chem.* **2009**, *57*, 5452–5460. [CrossRef]
38. Collivignarelli, M.C.; Abbà, A.; Miino, M.C.; Damiani, S. Treatments for color removal from wastewater: State of the art. *J. Environ. Manag.* **2019**, *236*, 727–745. [CrossRef] [PubMed]
39. Pakshirajan, K.; Kheria, S. Continuous treatment of coloured industry wastewater using immobilized *Phanerochaete chrysosporium* in a rotating biological contactor reactor. *J. Environ. Manag.* **2012**, *101*, 118–123. [CrossRef] [PubMed]
40. Ergül, F.E.; Sargın, S.; Öngen, G.; Sukan, F.V. Dephenolisation of olive mill wastewater using adapted *Trametes versicolor*. *Int. Biodeterior. Biodegradation* **2009**, *63*, 1–6. [CrossRef]

Article

Heterogeneous Fenton Oxidation with Natural Clay for Textile Levafix Dark Blue Dye Removal from Aqueous Effluent

Manasik M. Nour [1], Maha A. Tony [2,3] and Hossam A. Nabwey [1,2,*]

[1] Department of Mathematics, College of Science and Humanities in Al-Kharj, Prince Sattam Bin Abdulaziz University, Al-Kharj 11942, Saudi Arabia; h.mohamed@psau.edu.sa
[2] Basic Engineering Science Department, Faculty of Engineering, Menoufia University, Shebin El-Kom 32511, Egypt; dr.maha.tony@gmail.com
[3] Advanced Materials/Solar Energy and Environmental Sustainability (AMSEES) Laboratory, Faculty of Engineering, Menoufia University, Shebin El-Kom 32511, Egypt
* Correspondence: eng_hossam21@yahoo.com

Abstract: The ever-increasing technological advancement and industrialization are leading to a massive discharge of hazardous waste into the aquatic environment, calling on scientists and researchers to introduce environmentally benign solutions. In this regard, the current work is based on introducing Fuller's earth, which is regarded as an environmentally benign material, as an innovative Fenton oxidation technology to treat effluent loaded with Levafix Dark Blue dye. Initially, Fuller's earth was chemically and thermally activated, then subjected to characterization using a field-emission scanning electron microscope (FE-SEM) augmented with an energy-dispersive X-ray analyzer (EDX) and Fourier transform infrared (FTIR). This detailed the morphologies of the samples and the functional groups on the catalyst leading to the reaction with the dye. Fuller's earth, augmented with hydrogen peroxide, was then introduced as a photo-Fenton oxidation system under UV illumination for dye oxidation. Moreover, a response surface mythological analysis was applied to optimize the most effective operational parameters. The experimental data revealed that the optimal Fuller's earth dose corresponded to 1.02 mg/L using the optimal H_2O_2 of 818 mg/L at pH 3.0, and the removal efficiency reached 99%. Moreover, the thermodynamic parameters were investigated, and the data revealed the positive $\Delta G'$ and negative $\Delta S'$ values that reflect the non-spontaneous nature of oxidation at high temperatures. Additionally, the negative $\Delta H'$ values suggest the occurrence of the endothermic oxidation reaction. Furthermore, the reaction followed the second-order kinetic model. Finally, the catalyst stability was investigated, and reasonable removal efficiency was attained (73%) after the successive use of Fuller's earth reached six cyclic uses.

Keywords: wastewater; clay; Fuller's earth; Levafix Dark Blue dye; oxidation; Fenton

Citation: Nour, M.M.; Tony, M.A.; Nabwey, H.A. Heterogeneous Fenton Oxidation with Natural Clay for Textile Levafix Dark Blue Dye Removal from Aqueous Effluent. *Appl. Sci.* **2023**, *13*, 8948. https://doi.org/10.3390/app13158948

Academic Editors: Dae Sung Lee, Yolanda Patiño and Amanda Laca Pérez

Received: 19 May 2023
Revised: 21 July 2023
Accepted: 26 July 2023
Published: 3 August 2023

Copyright: © 2023 by the authors. Licensee MDPI, Basel, Switzerland. This article is an open access article distributed under the terms and conditions of the Creative Commons Attribution (CC BY) license (https://creativecommons.org/licenses/by/4.0/).

1. Introduction

Currently, since a greener environment is the hallmark of the scientific world, the contribution of naturally abundant clay minerals is notably visible. In recent decades, clay minerals based on Fuller's earth [1] have been used to modify the performance of photocatalytic reactions. Scientists are inspired by the outstanding characteristics of clay since it is abundant, cost-efficient, and benign to the environment and ecosystem [2]. Although Fuller's earth was first discovered in 1847, clay science is associated with prehistoric times. Its use in several applications can be traced back to 200 cultures, such as the ancient Egyptians, Amargosians, and South and North Americans [3,4]. Recently, Fuller's earth-based semiconductors have attracted great attention for their efficient performance in photocatalytic reactions for eliminating various contaminates from wastewater streams.

However, with the ever-increasing development of societies and the global industrialization revolution, the problems associated with environmental pollution are also tremendously increasing [5]. Massive amounts of organic dyes are discharged annually from various industries. Such industries include the printing and paper industry, photographic industries, tanning and leather industries, and textile dyeing industries [6]. The textile industry is considered the most polluting industry in the industrial sector [6]. This industry consumes substantial amounts of water in its processing and finishing. Thus, the result is a huge amount of wastewater contaminated with various dye species. Such waste causes persistent damage to the environment [7]. Most of the dyes included in this aqueous effluent are carcinogenic and cause severe damage to the ecosystem and its habitants. Hence, such contaminated wastewater must undergo treatment prior to its final disposal into the environment. When comparing photocatalytic reactions to other wastewater treatment technologies, the photocatalytic system is a superior candidate. This might be due to its complete mineralization tendency for contaminants in aqueous streams [8]. Among the photocatalytic reactions, the iron-based catalytic reaction that uses Fe^{2+} and Fe^{3+} as typical iron sources, the so-called Fenton reaction, has attracted scientists' attention for its unique photocatalytic activity [9]. Moreover, its use in contaminate elimination has increased due to the reaction being cost-efficient and the catalyst possessing optical properties [10]. Although the properties of this reaction are superior, there are three main defects that restrict its application. The obstructions are the chemical precursor's price, the need for an acidic pH medium for treatment, and the byproduct sludge after treatment that requires further handling prior to the final discharge [11]. From this concept, various trials have been developed to overcome these drawbacks of the Fenton process. For instance, introducing hetero-junction catalytic materials as the precursors of the Fenton reaction is an excellent strategy due to both its superior catalytic advances and its recovery ability [12]. Moreover, replacing Fe^{2+} or Fe^{3+} with non-iron metals, such as Cu^{2+}, shows excellent results since this widens the acidic pH range. Furthermore, aluminum has been shown to be an excellent replacement for Fe^{2+} or Fe^{3+} in the reagent as anon-iron Fenton system [13]. Moreover, the aluminum that is present in natural-based materials [14] might be applied to initiate the iron-based Fenton system as a non-iron system [15].

Scientists' crucial goal is identifying environmentally benign materials. In this regard, the naturally abundant Fuller's earth is a suitable candidate. Fuller's earth is the main component of clay minerals that comprise silicon, calcium, and aluminum oxides with a dominant fraction of Al_2O_3. The elementary molecular structure is an aluminum octahedral structure [3]. Due to its environmental benignity, Fuller's earth is applied in the fields of drug delivery and wastewater treatment. The recently published volume of literature associated with Fuller's earth and its applications indicates an interest in using Fuller's earth in various applications, especially in wastewater treatment. Such wastewater management applications include adsorption techniques or augmentation with semiconductors to act as a photocatalyst. For instance, Safwat et al. [5] used Fuller's earth augmented with kaolin for the elimination of phenolic compounds from an aqueous stream via adsorption methodology. Moreover, Shah and his co-workers [9] used a modified form of Fuller's earth in combination with a surfactant to improve its adsorption capacity for eliminating acid red 17 dye from wastewater. However, to the best of the authors' knowledge, it has not been applied in its solo form as a photocatalyst for dye removal, especially as a source of the Fenton reaction. Catalysts from raw clay, such as Fuller's earth, could represent environmentally friendly and environmentally benign catalysts that possess many advantages. Such materials are cost-efficient and naturally abundant, in addition to being non-toxic to the environment, and they also possess excellent properties, such as a high surface area.

Fuller's earth clay is critical for creating ˙OH radical species, and this is achieved by using the elements in Fuller's earth, such as aluminum and iron ions [14]. Such ions react with hydrogen peroxide, and the reaction is initiated by ultraviolet light and then produces hydroxyl radicals. Iron and aluminum ions are formed and react with hydrogen peroxide to form further hydroxyl radicals and elemental ions (Equations (1) and (2)). ˙OH radicals are categorized as non-selective species that attack pollutant molecules and strongly oxidize them. Aluminum might initiate the Fenton reaction via the acyclic reaction [15].

A general trend of the aluminum-based Fenton reaction mechanism is the formation of the Al^{3+} superoxide complex (Equation (3)) [16]. In this reaction, the aluminum is able to stabilize a superoxide radical (O_2^{-} anion), and thus, the formed Al^{3+} superoxide complex (Equation (4)) is capable of reducing Fe^{3+} to Fe^{2+}. Hence, Fe^{2+} could enhance the production of ·OH radicals through the Fenton reaction (Equation (5)) [17,18].

$$Fe^{2+} + H_2O_2 \rightarrow Fe^{3+} + OH^{-} + OH^{·} \quad (1)$$

$$Fe^{3+} + H_2O_2 \rightarrow Fe^{2+} + OH_2^{·} + H^{+} \quad (2)$$

$$Al(H_2O)_6^{3+} + O_2^{-} \rightarrow Al(O_2)(H_2O)_5^{2+} + H_2O \quad (3)$$

$$Al(O_2)(H_2O)_5^{2+} + O_2^{-} \rightarrow Al(O_2)(OH)(H_2O)_4^{+} + H_2O \quad (4)$$

$$Fe^{3+} + AlO_2^{2+} \rightarrow Fe^{2+} + AlO_2^{3+} \quad (5)$$

To the best of the authors' knowledge, according to the cited literature, "Fuller's earth" has not yet been applied as an oxidation source for pollutant remediation. Traditionally, clay sources are applied as adsorbent materials. Therefore, this investigative study introduces the novel application of Fuller's earth as the elemental source of Fenton oxidation. The goal of the current work is based on altering the traditional Fenton source with the environmentally benign, naturally available, and abundant clay "Fuller's earth", which comprises various metals. Such metals lead to Fenton reaction oxidation. The system is applied to mineralize Levafix Dark Blue aqueous effluent as a simulation of textile-effluent-polluted wastewater. The influence of various operating variables, i.e., Fuller's earth and the hydrogen peroxide reagent concentration, the pH of the medium, dye loading, and the temperature of the wastewater are assessed in order to meet the real application requirements.

2. Materials and Methods

2.1. Wastewater

Levafix Dark Blue dye, which is a kind of reactive azo dye, is commonly applied in the textile dyeing industry due to its high fastness profile, and it meets most requirements set by textile manufacturers. In this regard, commercial Levafix Dark Blue dye was used in the current study as a model synthetic pollutant. Thus, Levafix Dark Blue dye was used to prepare synthetic wastewater effluent. Levafix Dark Blue was supplied by DyStar Management Co., Ltd., Shanghai, China. The dye was used as received without further purification or treatment. The dye is dark blue and regarded as a bi-functional, combined anchor. The dye powder was used with no purification or further treatment. Initially, to attain the synthetic dye effluent, a stock solution of 1000 ppm of Levafix Dark Blue dye was prepared, which was then diluted, as required, for successive dilutions to obtain different concentrations according to the experimental conditions.

2.2. Preparation of Fuller's Earth-Based Fenton Catalyst

Naturally occurring Fuller's earth clay was collected from a deposit located in the southeastern desert in Egypt. After collection, the Fuller's earth clay was subjected to electric oven drying (105 °C) to remove any moisture content. Generally, the characteristics of clay, including its chemical and physical features, might be modified and enhanced through various treatments in order to improve its natural capacity to achieve better treatment results [14]. Such modifications could improve the physicochemical and mineralogical characteristics of the substance. Acid and thermal treatments of Fuller's earth are techniques widely applied to attain clay modification. Such techniques modify the mineralogical composition and chemical structure of Fuller's earth substance, as well as leading to surface activation according to the authors' preliminary work. Thus, next, the Fuller's earth was ball-milled to attain a fine powder. Afterward, the material was sieved (200 mesh) and cooked with hydrochloric acid. Then, 15 gm of the material was cooked with 200 mL HCl acid (10 M) through heating (70 °C) and stirring for 1.5 h. Subsequently, the resultant aqueous media were successively washed with distilled water to reach a neutral pH. Subsequently, the solution was filtered, and the resultant solid powder was subjected to calcination (600 °C) for thermal activation purposes.

2.3. Photocatalytic Test

A stock solution of 1000 ppm was prepared from Levafix Dark Blue, and a further dilution was carried out when needed to obtain 50, 100, 150, and 200 ppm. Initially, 100 mL of the 100 ppm of Levafix Dark Blue dye-containing aqueous solution was poured into a glass container to subject it to the photocatalytic test. Then, a certain amount of Fuller's earth augmented with hydrogen peroxide (30% w/v) supplied by Sigma-Aldrich (Burlington, MA, USA) was poured into a container as the source of the Fenton photocatalyst and placed in a photochemical reactor. The mixture was magnetically stirred and kept under UV illumination after the pH was adjusted, if desired (using AD1030, Adwa instrument, Szeged, Hungary), over the range of 3.0 to 8.0. The pH of the dye aqueous solution was adjusted to the desired values by using diluted H_2SO_4 (1:9) and/or 1M NaOH solutions (Sigma-Aldrich). All chemicals were used as received from the supplier without further treatment or purification.

In order to validate the effect of the Levafix Dark Blue dye concentration on the extent of photocatalytic oxidation, the polluted water with the reagents was subjected to the photocatalytic system. A UV lamp (15 W, 230 V/50 Hz, with a 253.7 nm wavelength) was used to emit UV light during the reaction. The lamp was covered with a silica tube jacket for lamp protection, still allowing the UV to penetrate the dye-containing solution. The sleeved UV lamp was located inside a glass vessel containing the wastewater solution to well induce and accelerate the photocatalytic reaction. The photo-reactor had a 250 mL volume, and the reactor was fully exposed to UV light.

In regular time intervals (every 10 min), the samples were subjected to analysis after filtration (0.45 µm) to remove the remaining excess catalyst using a UV–visible spectrophotometer (Unico UV-2100, Franksville, WI, USA). The results of the analysis were recorded, and the data are presented as the dye percentage removal. The experimental setup is summarized in Figure 1.

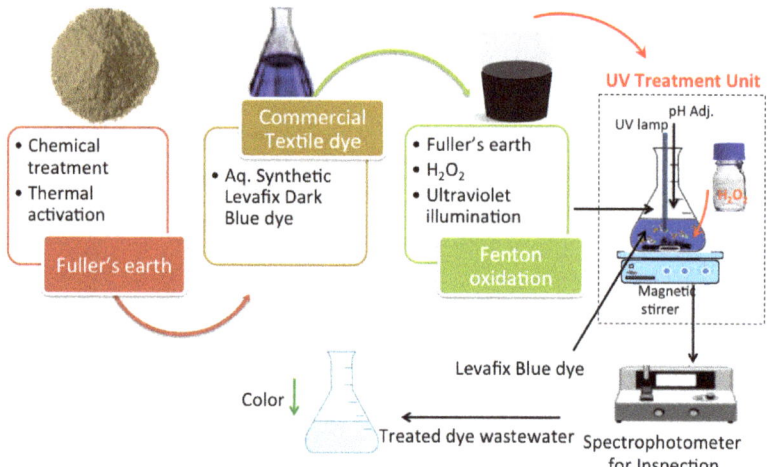

Figure 1. Schematic representation of the catalyst preparation and treatment steps.

2.4. Characterization Study

The morphologies of the attained Fuller's earth sample were explored and imaged using a field-emission scanning electron microscope (FE-SEM) (FE-SEM, Quanta FEG 250,

Technical Cell, Kolkata, India). The typically used magnifications were ×8000 and ×60,000. Furthermore, this instrument was supplemented by energy-dispersive X-ray spectroscopy (EDX) in order to assess the content of principal oxides in the Fuller's earth. The oxides were examined via the energy-dispersive spectrum. Moreover, a Fourier transform infrared FTIR spectrum was obtained using the Jasco FT/IR-4100 model type (Jasco Inc., Mary's Court Easton, MD, USA).

2.5. Statistical Analysis

To attain a further higher reasonable dye removal efficiency with a good understanding of the roles of the significant independent parameters in the Levafix dye oxidation, a response investigated via dye percentage as dependent variable removal, a three-level factorial design, that is, the so-called Box–Behnken design, with triplicates of the central points, was applied. The selected independent variables were chosen according to the most affecting factors as follows: (i) H_2O_2 dose; (ii) Fuller's earth doses; and (iii) pH value. The levels and range for each parameter were selected according to a preliminary study that determined the manually optimized values. The selected levels and ranges are displayed in Table 1. Subsequently, SAS, statistical analysis software (SAS, Institute USA, Cary, NC, USA), was used to propose the full factorial experimental design matrix. Then, Matlab (7.11.0.584) software, as well as an analysis of variance (ANOVA), was chosen in order to analyze and specify the significance of the statistical technique. Moreover, Mathematica (V 5.2) software was selected in order to determine the optimal values. Finally, extra triplicates of the experiments were conducted to validate the proposed investigated model equation.

Table 1. Boundaries of the uncoded and coded experimental domains of the Box–Behnken factorial design with respect to their level spacing.

Experimental Variables	Symbols		Range and Levels		
	Uncoded	Coded	−1	0	1
H_2O_2 (mg/L)	ε_1	ζ_1	700	800	900
Fuller's earth (mg/L)	ε_2	ζ_2	0.75	1.0	1.25
pH	ε_3	ζ_3	2.5	3.0	3.5

3. Results and Discussion

3.1. Characterization of the Prepared Fuller's Earth Material

Scanning electron microscopy (SEM) was used to explain the treated and calcinated morphologies of the Fuller's earth material; an image is displayed in Figure 2. The surface was observed to have an irregular shape, and the clay possessed lots of asymmetric, open pores. The voids aid adsorption, as well as catalytic oxidation activity, a source of various metals. Moreover, elemental analysis techniques of the Fuller's earth material using an energy-dispersive X-ray analyzer (EDX) revealed its chemical composition, which constituted different oxides. The predominant oxide in the Fuller's earth was SiO_2, as well as Al_2O_3 and Fe_2O_3. The presence of Al_2O_3 and Fe_2O_3 leads to the occurrence of the Fenton reaction. Furthermore, trace amounts of CaO, MgO, K_2O, SO_3, and Na_2O were present in the clay. A previous study [15] confirmed that the presence of Al_2O_3 leads to the Fenton reaction [18]. Moreover, various studies [15–18] verified that such oxides improve the adsorption tendency of Fuller's earth clay.

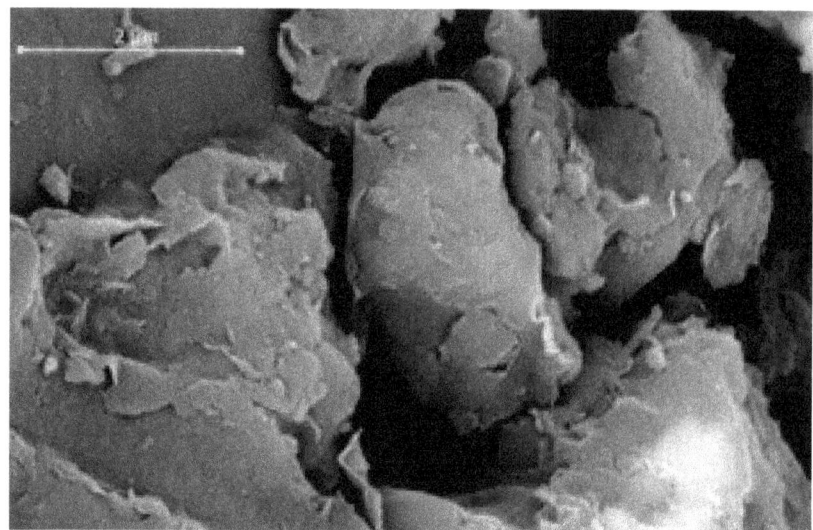

Figure 2. FE-SEM micrograph image of the prepared catalyst "Fuller's earth clay".

An oxide material analysis of the Fuller's earth clay data obtained via an EDX analysis is displayed in Table 2. The main oxide of the chemically treated and then calcined modified Fuller's earth powder was SiO_2, along with Al_2O_3 and Fe_2O_3, with the presence of small amounts of the oxides Ca, Mg, K, and Na. The use of Fuller's earth, which is considered a clay, is due to its composition.

Table 2. Chemical oxide composition of modified Fuller's earth determined via EDX.

Element	SiO_2	Al_2O_3	Fe_2O_3	CaO	MgO	K_2O	Na_2O	LOI
weight %	79.16	12.81	5.26	0.53	0.96	0.08	0.24	0.96

LOI: Loss of Ignition.

Although previous data [19] have indicated the importance of the oxides SiO_2, Al_2O_3, and Fe_2O_3 in enhancing the adsorption capacity of pollutants, Al_2O_3 and Fe_2O_3 are sources of the Fenton reaction [15–18]. Such oxides might be initiated through hydrogen peroxide to generate hydroxyl radicals, which are categorized as highly reactive intermediates and possess the ability to oxidize pollutant species [13].

A Fourier transform infrared (FTIR) transmittance spectrum analysis was applied as a suggestive technique. FTIR was applied to identify the different forms of minerals present in the Fuller's earth that were introduced as a source for the Fenton oxidation test. Moreover, FTIR provided information on the chemical nature of the Fuller's earth substance, as well as some details about the interactions. As illustrated in Figure 3, the FTIR performances of the Fuller's earth clay exhibited coupled vibrations that are mainly due to the existence of numerous elements. Additionally, the main absorption-intensive bands of Fuller's earth clay appeared. Silanol, which is characterized by Si–O stretching vibrations, is located at the 1034 cm^{-1} wavenumber. The occurrence of silanol represents the existence of quartz. Moreover, the bands at 528.4 cm^{-1} and 779 cm^{-1} are associated with the presence of illite, which is reflected by the Si-O-Al group. Moreover, the interlayer hydrogen bonding illustrates the possibility of hydroxyl linkage. Al-Mg-OH and Si-O-Fe bonding are reflected by the bands at 779 and 820 cm^{-1}, respectively. This confirmed the existence of Fe and Al bonding, which is necessary for the Fenton oxidation test.

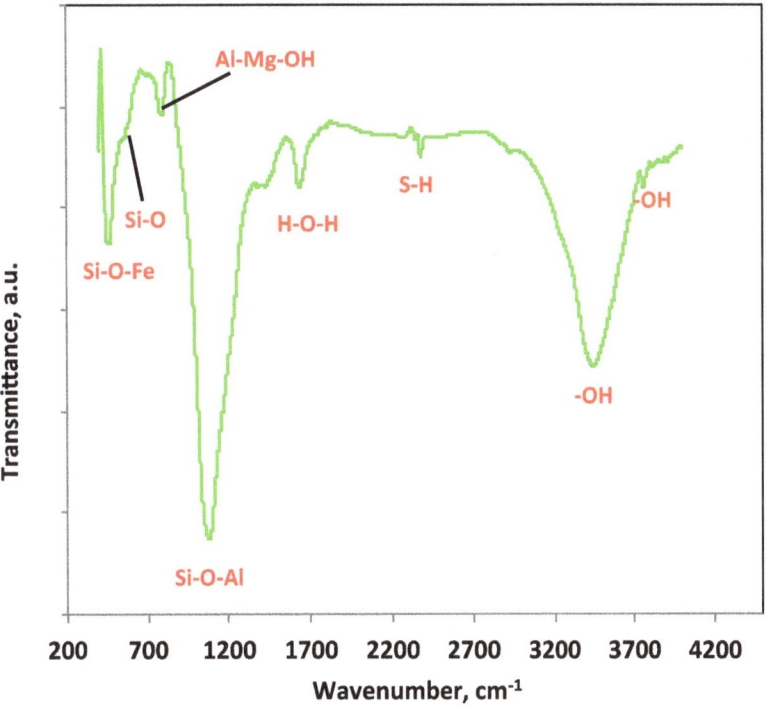

Figure 3. FTIR spectrum of the Fuller's earth clay.

3.2. Reactive Azo Dye Oxidation

3.2.1. Effects of Reaction Time and Reactive Azo Dye Loading

Each of the experimental parameters affecting the kinetics of the reaction was investigated, starting with the experimental reaction time. In this experiment, the influence of the oxidation system was examined in terms of the photo-Fenton oxidation reaction based on Fuller's earth natural clay. In order to examine the optimal reaction time and its impact on the oxidation process, experiments were undertaken with illumination times ranging from 2 to 60 min. All other parameters, including the initial concentrations of Fuller's earth and H_2O_2 at 1.0 g/L and 800 mg/L, respectively, were kept constant, and the solution pH was maintained at 3.0. Figure 4 demonstrates the influence of the reaction time on the profiles of various reactive azo dye loadings. An investigation of the results in Figure 4 indicated that the dye oxidation rate reached 73% within only the initial two minutes of the illumination time. However, afterward, it steadily declined, achieving an accumulative oxidation efficacy reaching 97% within 15 min when the initial azo dye concentration was 50 mg/L. According to the previously cited literature [16,20], azo dye molecules compromise aromatic rings. The hydroxyl radicals generated from the reaction of Fuller's earth with hydrogen peroxide attack such rings and open them. Then, this generates reaction intermediates and eventually oxidizes them to harmless end products (CO_2 and H_2O). Initial rapid oxidation was dominant for all the dye concentrations. As the reaction proceeded, the generated hydroxyl radicals in the reaction medium gradually declined, corresponding to a reduction in the H_2O_2 concentration. Moreover, radicals other than hydroxyl radicals were produced, and they inhibited the oxidation reaction rate rather than improving the dye removal rate. The previous conclusion by [13] confirmed this investigation by treating various dyes contaminating wastewater effluent.

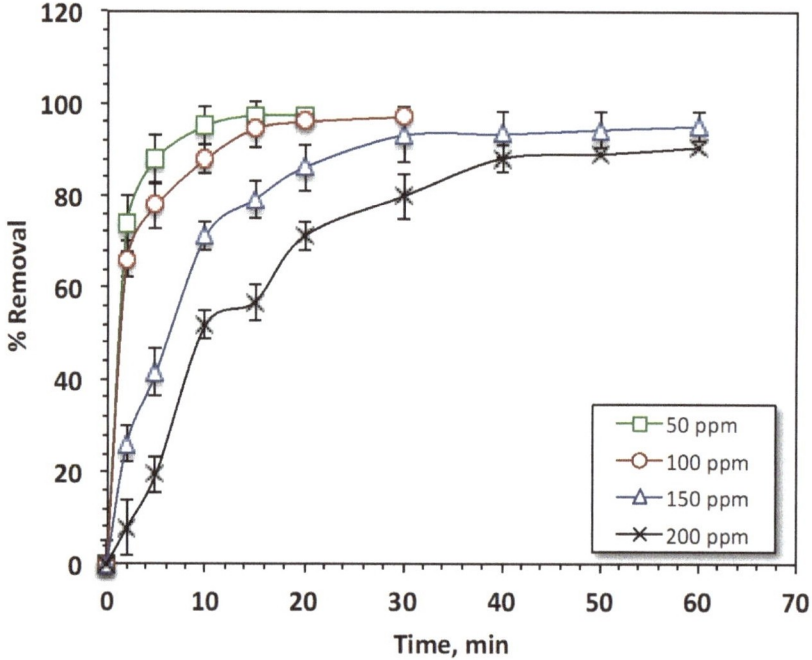

Figure 4. Effects of azo dye Levafix Dark Blue loadings on the oxidation reaction (experimental conditions: Fuller's earth 1.0 g/L; H_2O_2 800 mg/L; and pH 3.0).

While a similar rapid reaction rate in the initial reaction time period was observed for all the treated dye concentrations, an assessment of Figure 4 found that the oxidation rate declined with the increase in the dye loading. The removal efficiencies were 97, 95, 91, and 79% for the 50, 100, 150, and 200 mg/L azo Levafix Dark dye concentrations in the aqueous effluent, respectively. Moreover, the reaction time also increased from 20 to 100 min with the increase in the dye loading. At higher dye concentrations, the concentration of the exerted ˙OH radicals was insufficient for completing dye oxidation. This conclusion of increasing the oxidation activity with a reduction in the initial pollutant concentration was also previously investigated and recorded by Raut-Jadhav et al. [21], who used the Fenton reagent for the oxidation of methomyl-pesticide-containing wastewater.

3.2.2. Effects of Various Treatment Systems

The effects of various treatments based on oxidation systems, namely, Dark/Fuller earth, Dark/H_2O_2, Dark/H_2O_2-Fuller earth, UV/H_2O_2, UV/Fuller earth, and UV/H_2O_2-Fuller earth, were explored to test the effect of the Fenton reaction tendency using Fuller's earth as a catalyst. The efficacy of these techniques was assessed in terms of Levafix Dark Blue dye color removal, and the results are exhibited in Figure 5. The data compare the Levafix Dark Blue oxidation using the different systems at room temperature. It is clear from the data in Figure 5 that the solo H_2O_2 oxidation system results in the dye having oxidation efficiencies of only 7% and 38% in the dark and under ultraviolet illumination, respectively, within 20 min of the reaction time. Moreover, in the dark test and under UV illumination, the solo Fuller's earth catalyst reached removal efficiencies of 26 and 39%, respectively. However, under the solo UV illumination system, the dye oxidation reached a removal rate of only 11%. In contrast, the augmented Fuller's earth with hydrogen peroxide oxidative treatment could mineralize the dye to attain rates of 72 and 98% in the dark and under UV illumination, respectively. It is worth mentioning that the photo-Fenton system based on Fuller's earth augmented with hydrogen peroxide achieved the highest oxidation among the investigated oxidation systems. Notably, the high removal rates of the Fenton system or the photo-Fenton system compared to those of the solo hydrogen peroxide systems or the Fuller's earth systems verify the role of the Fenton reaction in eliminating dye molecules. The presence of dual reagents might explain this in the Fenton system, as they support the generation of more hydroxyl radicals. Moreover, the presence of ultraviolet light results in further hydroxyl radical production, which is the main responsibility of the oxidation system. This reaction trend was previously reported by Thabet et al. [13], who treated contaminated dye effluent using a modified Fenton system.

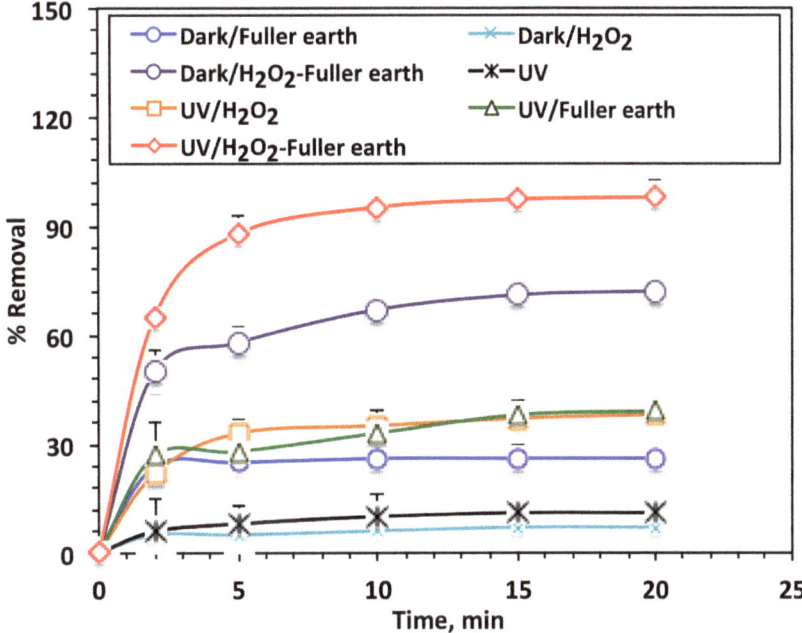

Figure 5. Effects of different treatment systems on azo dye Levafix Dark Blue (experimental conditions: azo dye Levafix Dark Blue 50 ppm; Fuller's earth 1.0 g/L; H_2O_2 800 mg/L; and pH 3.0).

However, with an increase in time, the Levafix Blue dye removal exhibited a slower rate since ˙OH radical production declined. Contrary, a rapid initial stage of oxidation was achieved. This is due to the consumption of the reagent in the initial stage to produce hydroxyl radicals [22]. The oxidation occurred by utilizing the activation of H_2O_2 with the metal salts present in Fuller's earth in the case of the Fenton system. However, by exceeding the reaction time, a reduction in the dye rate was attained that corresponded to the decline in the presence of H_2O_2. This is related to the hydroxyl radical formation. Moreover, this phenomenon is also exhibited in hydrogen peroxide-based systems since the amount of hydrogen peroxide reduces due to it being consumed as time proceeds. Various research data [23] have exhibited that the quick initial reaction time is due to the immediate generation of ˙OH radical species. Furthermore, the lower oxidation tendency of the dark reaction test is associated with the absence of UV illumination. Thus, it initiated oxidation through the generation of more ˙OH radical species. Remarkably, the Fuller's earth catalyst is easily available since it is naturally abundant as a sustainable substance.

3.2.3. Effects of Photo-Fenton Parameters
Effect of Fuller's Earth Dose

To examine the effect of the Fuller's earth catalyst, as a source of the Fenton reaction, on the Fenton oxidation of azo dye, experiments were undertaken to investigate the influence of clay dose on the reaction kinetics. Figure 6 displays its influence on the oxidation reaction by varying the Fuller's earth concentration over the range of 0.5 to 1.5 mg/L.

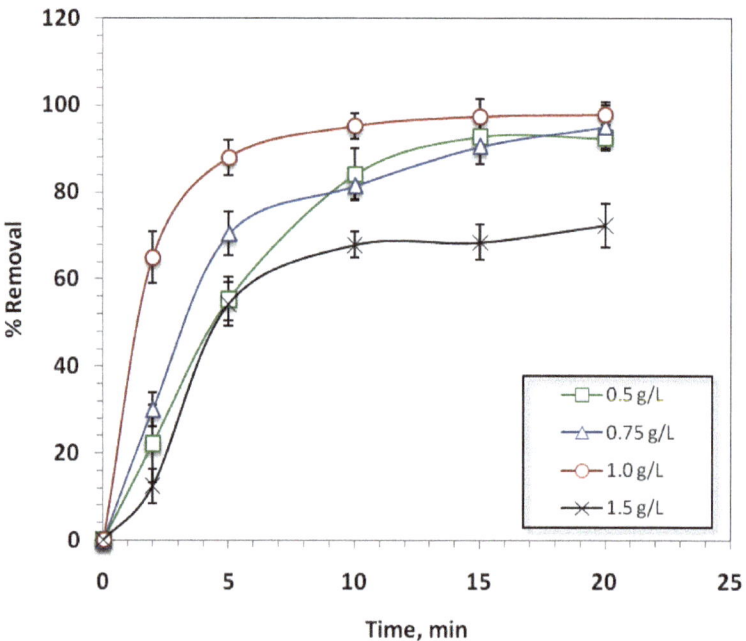

Figure 6. Effect of Fuller's earth concentration on photo-Fenton oxidation system (experimental conditions: azo dye Levafix Dark Blue 50 ppm; H_2O_2 800 mg/L; and pH 3.0).

The oxidation removal efficacy was enhanced (from 92% to 97%) by increasing the Fuller's earth dose from 0.5 to 1.0 mg/L. However, further increasing Fuller's earth to 1.5 mg/L resulted in a significant decline in dye oxidation, reaching only 72%. This could be associated with the Al and Fe ions that form in the aqueous solution since they are present in Fuller's earth and then react with H_2O_2 to form further ˙OH radicals and metal ions. These non-selective ˙OH radicals attack the dye molecules and strongly oxidize them. However, a further increase in the Fuller's earth catalyst resulted in a decline in the oxidation efficiency as a result of the shadowing effect arising from the turbid solution. This phenomenon of shadowing the UV illumination is due to the Fuller's earth comprising inorganic materials that obey UV transmittance when it is in excess. Additionally, an excessive concentration of metal ions results in their acting as ˙OH radical scavengers rather than generators according to the abovementioned equations (Equation (5)). This investigation is in accordance with the studies in [24–28], in which organic materials and dyes from wastewater are treated via the Fenton reaction.

Effect of Hydrogen Peroxide Reagent Concentration

Figure 7 displays the reactive dye Levafix Dark Blue oxidation in the presence of the Fuller's earth-based photo-Fenton system. The oxidation rates elevated when the H_2O_2 reagent dose was increased from 200 to 800 mg/L, until reaching practically 97% Levafix Dark oxidation. This occurred with a 20 min illumination time using the optimal value of 800 mg/L of H_2O_2, 1 g/L of Fuller's earth, and a solution pH of 3.0. This increase in the removal efficiency with the increase in the reagent is associated with the production of more peroxide species and hydroxyl radicals in the aqueous medium. These radicals are the horsepower of the oxidation reaction, as well as its main responsibility. Moreover, elevating the H_2O_2 concentration to 1000 mg/L, more than the optimal limit (800 mg/L), affects the reaction rate and results in a decline in oxidation, reaching only 74%. This is probably related to the excess reagent concentration of ion pairs on the surface that could reduce the availability of the Fuller's earth elemental sites to react with the excess H_2O_2 (Equations (6) and (7)). In such a situation, additional H_2O_2 may have a detrimental effect due to the scavenging of hydroxyl radicals that occurs at higher H_2O [24]. Moreover, there are radicals other than OH radicals in the presence of excess hydrogen peroxide; such radicals inhibit the oxidation reaction (Equations (8) and (9)).

$$H_2O_2 + h\upsilon \rightarrow 2OH^* \tag{6}$$

$$OH^* + OH^* \rightarrow H_2O_2 \tag{7}$$

$$H_2O_2 + OH^* \rightarrow H_2O + OOH^* \tag{8}$$

$$OOH^* + OH^* \rightarrow H_2O + O_2 \tag{9}$$

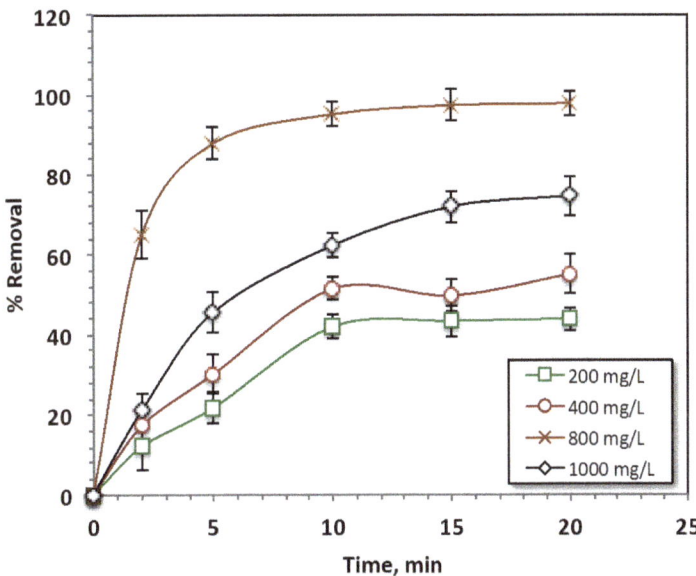

Figure 7. Effect of hydrogen peroxide concentration on photo-Fenton oxidation system (experimental conditions: azo dye Levafix Dark Blue 50 ppm; Fuller's earth. 1.0 g/L; and pH 3.0).

Effect of pH Value on the Treatment Efficiency

The pH of the aqueous environment is categorized as a vital influencing variable in Fenton oxidation. This demonstrates its importance since the pH influences H_2O_2 decomposition and the hydrolytic speciation of metal ions. In this regard, to evaluate its influence on the Levafix Dye oxidation through the modified photo-Fenton system, the initial pH values were altered from an acidic value to an alkaline value to evaluate their effects on the system. The pH values varied over the range of 3.0 to 8.0 according to the results displayed in Figure 8; an alkaline pH value is not favorable for the oxidation reaction using the photo-Fenton system. However, further oxidation was achieved when the acidic pH was used. Notably, it was obvious that the acidic pH (3.0) value of wastewater attained the highest removal efficiency. The removal efficiency reached 97% within 20 min of the illumination time, at which point the Levafix dye oxidized into other intermediates. However, increasing the aqueous solution pH to an alkaline pH value results in the presence of excess unfavorable ions. Such ions react with superoxide species rather than the metals in the catalyst. This hinders the overall reaction. These metals are mostly accountable for provoking H_2O_2 to produce (OH) radicals [27]. The result is the formation of the hydroperoxide (HO_2) inactive radical, which reduces the oxidation efficacy [28]. Hence, the Levafix oxidation yield further declined at an alkaline pH. The findings of this investigation of the low efficiency at a high pH are in accordance with the previous findings of Nichela et al. [15], who treated a nitrobenzene-contaminated aqueous stream using a Fenton-based reaction.

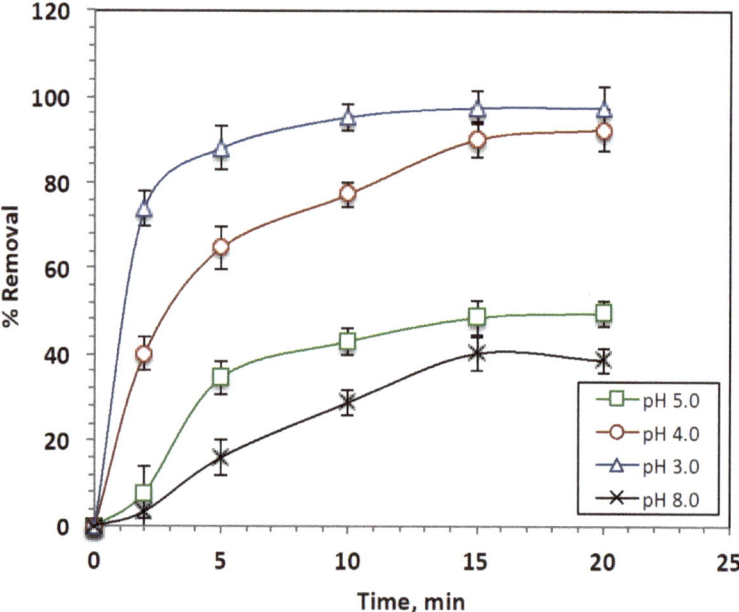

Figure 8. Effect of pH value on photo-Fenton oxidation system (experimental conditions: azo dye Levafix Dark Blue 50 ppm; Fuller's earth 1.0 g/L; and H_2O_2 800 mg/L).

Box–Behnken Regression Design Fitting

RSM, response surface regression methodology, was applied to explore the operational parameters' optimal values for the modified photo-Fenton system based on the Fuller's earth clay system. These parameters include the Fuller's earth clay and hydrogen peroxide concentrations, as well as the pH value, in order to maximize the Levafix dye efficacy removal. The used outlined matrix is tabulated in Table 1, as well as the experimental dye response after the oxidation reaction, with the predictive model values presented using Equation (10). This equation exploring the second-order polynomial regression model exhibits the response surface in terms of the coded variables for the Levafix dye removal response.

$$Y(\%) = 90.47 + 2.28\,\varepsilon_1 - 4.68\,5\varepsilon_2 - 5.08\,\varepsilon_3 - 4.75\,6\varepsilon_1^2 + 0.23\,\varepsilon_1\varepsilon_2 - 12.57\,\varepsilon_1\varepsilon_3 - 4.98\,\varepsilon_2^2 + 6.3\,\varepsilon_2\varepsilon_3 - 18.41\,\varepsilon_3^2 \tag{10}$$

An ANOVA test based on Fisher's statistical analysis was performed for the assessment of the statistical consequence and the adequacy of a quadratic model. With a minimum deviation, a small probability value (<0.005), and a high R^2 (the regression coefficient value), the model was shown to be the best fit. The R^2 value of the Levafix Dark dye oxidation response was 94.3%.

A graphical representation of the abovementioned equation (Equation (10)) demonstrates the influences of the experimental parameters on the response. The 3-D (three-dimensional) surface and the 2-D (two-dimensional) contour plots of the operational parameters were designed using Matlab software, and the results are displayed in Figure 9a–c. The data displayed in the figure revealed the response of each experimental influencing variable and the major interactions between those variables. An inspection of the 3-D surface graph and the 2-D contour plot in Figure 9b illustrated that the removal rate of the Levafix dye was enhanced with the increase in the concentrations of both H_2O_2 and Fuller's earth. However, the curvature displayed in Figure 9a indicates that there is a significant interaction effect between the Fuller's earth and hydrogen peroxide doses. This interaction encourages the generation of hydroxyl radical species that have a positive effect on the dye removal rate. Nevertheless, a further increase in the concentrations of both reagents resulted in a reduction in the dye oxidation rate. Thus, an optimal ratio of Fuller's earth/H_2O_2 is essential to increase the yield of ($^.$OH) radicals [27].

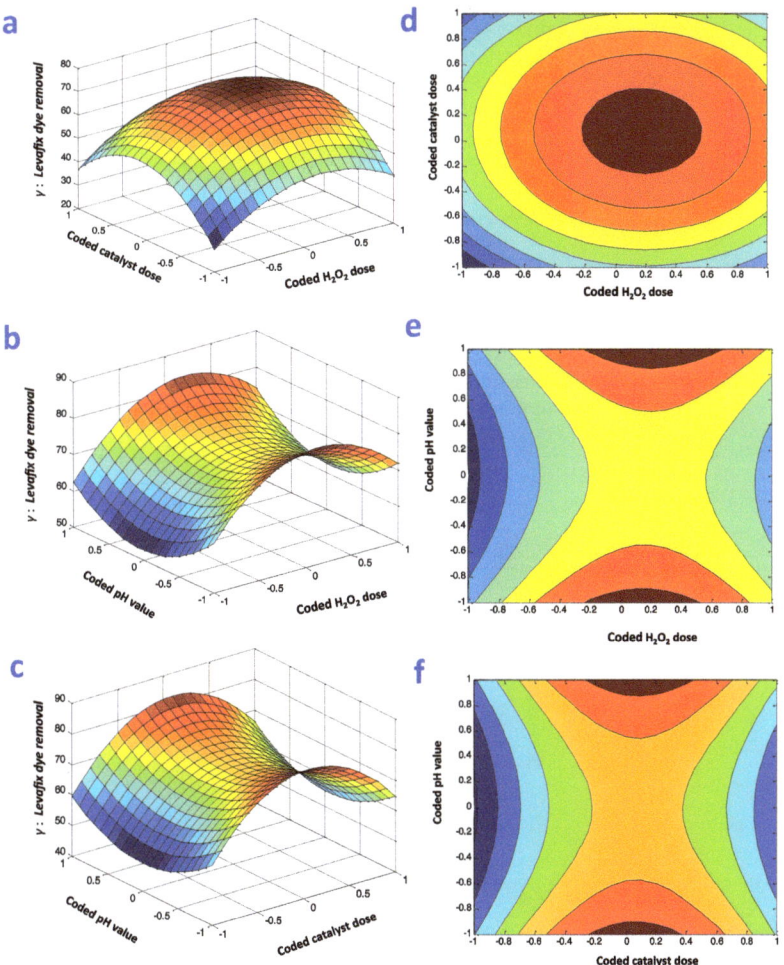

Figure 9. The 3-D (**a**–**c**) and 2-D (**d**–**f**) surface and contour plots of the coded independent variables and their responses in Levafix Dye removal: (**a**) H_2O_2 and Fuller's earth catalyst concentrations; (**b**) H_2O_2 concentration and pH; (**c**) Fuller's earth catalyst concentrations and pH.

Figure 9b explains the influence of the Fuller's earth catalyst dose and the significant parameter of pH in the augmented 3-D and 2-D response surface and contour plots. The graph confirms that the oxidation efficiency enhanced with an increasing catalyst dose. However, an alkaline environment is not favorable for Levafix oxidation. The optimum pH is near the acidic pH 3.0. The existence of scavenger species is apparent in the reaction medium rather than the reactive hydroxyl radicals in the reaction media. So, the result is a reduction in the overall reaction of oxidation, clarifying this finding.

The combined 3-D and 2-D response surface and contour plots displayed in Figure 9c demonstrate the influences of the independent hydrogen peroxide and pH values, respectively, on the Levafix dye as azo dye removal rates. An investigation was carried out on the figure results of Levafix Dye oxidation. In this set of experiments, the pH varied over the range of 2.5–3.5. The surface and contour plots of the RSM show that the highest percentage of dye removal was obtained at the intersection near the origin of the two variables.

For an extra examination of the proposed model, the statistically optimized predicted parameters investigated using Mathematica software were examined. The optimal conditions revealed from the predicted model were 818 mg/L, 1.02 mg/L, and 3.0 for H_2O_2, Fuller's earth catalyst, and pH, respectively. Then, an extra three replicates of experiments were carried out to confirm the predicted values' adequacy. After a 2 min reaction time, the measured percentage of dye removal (71%) was close to the predicted value (72%), which was obtained by applying the suggestive factorial design. Moreover, the overall oxidation reaction reached a 99% dye removal rate within 15 min of the reaction time. This investigation verifies that the response surface methodology is a satisfactory approach for optimizing the operational parameters' influence using the Fuller's earth-based Fenton system.

3.2.4. Effect of Temperature

In catalytic illumination oxidative systems, temperature is considered a vital parameter that affects reaction rates. To explore the influence of temperature on the reaction kinetics, Levafix Dark Blue dye oxidation experiments over a temperature range from 25 °C to 60 °C were undertaken. The results in Figure 10 display superior oxidation leading to an increase in the Levafix Dye removal efficacy, achieving complete removal with a shorter reaction time when the temperature is elevated.

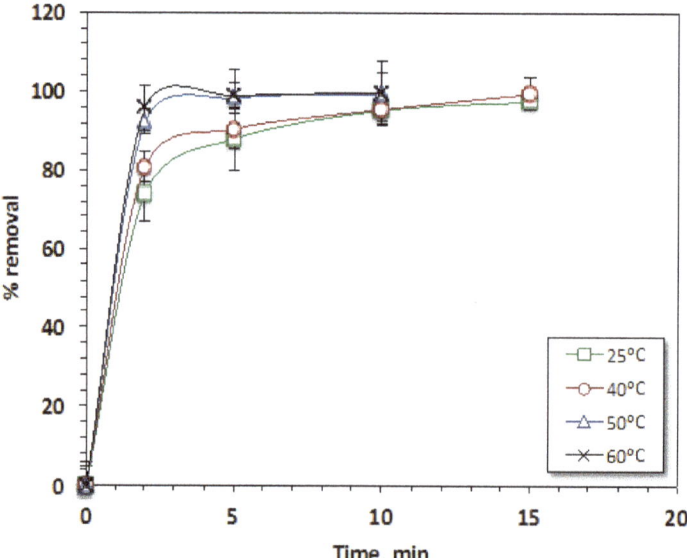

Figure 10. Temperature effect on Levafix removal via Fullers' earth-based Fenton process.

An examination of Figure 10 found that elevating the temperature from 25 °C to 60 °C for the aqueous Levafix solution attained a shorter reaction time of only 10 min and an enhancement from 97% to 100%. According to the previously cited literature [24], reaction rates are normally more efficient at higher temperatures. The various data cited in the literature [29,30] show that temperature elevation has a positive effect on Fenton systems for some wastewater-treated effluents.

3.2.5. Kinetics and Thermodynamics

In this section, the reaction kinetics and thermodynamics of Levafix dye oxidation via the photo-Fenton-modified system are investigated. In the current study, the photo-Fenton oxidation kinetics of Levafix dye using a modified reaction, the Fuller's earth-based photo-Fenton process, was evaluated for various contact times varying from 0 to 20 min under isothermal conditions at the following operating temperatures: 25, 40, 50, and 60 °C. Moreover, the zero-, first-, and second-order reaction kinetics were assessed for the modified Fenton oxidation reaction according to the following equations: Equation (11) for the zero-, Equation (12) for the first-, and Equation (13) for the second-order reaction kinetics [31]:

$$C_t = C_o - k_o t \tag{11}$$

$$C_t = C_o - e^{k_1 t} \tag{12}$$

$$\left(\frac{1}{C_t}\right) = \left(\frac{1}{C_0}\right) - k_2 t \tag{13}$$

where C is the concentration of the Levafix dye; C_t is the concentration of the Levafix dye at time t; C_o is the Levafix dye's initial concentration; t is the reaction time; and k_0, k_1, and k_2 represent the kinetic rate constants for the zero-, first-, and second-order reaction kinetics, respectively.

The reaction kinetics most appropriate for Levafix removal were assessed by plotting Equations (11)–(13) for the experimental results data. The kinetic parameters, as well as the regression coefficients (R^2), for each reaction order were investigated, and the data are tabulated in Table 3. Examining the data in Table 3 revealed that the reaction is well-fitted to a second-order reaction. Further, the kinetics constant of the second-order reaction constant, k_2, was notably affected by the reaction temperature, increasing with the temperature elevation. This investigation is associated with the generation of the ˙OH species since it is a product of the reaction of Fuller's earth with hydrogen peroxide. Moreover, another kinetics value of importance is $t_{1/2}$ (the half-life of a reaction), which signifies the essential time needed for the reactant's initial concentration to decrease by half. An examination of Table 3 revealed that the calculated $t_{1/2}$ is a function of the reaction temperature; $t_{1/2}$ declines with increasing temperature. Various researchers in other studies have confirmed that the Fenton reaction follows second-order reaction kinetics.

To fully understand the modified Fenton reaction based on the use of Fullers' earth oxidation to oxidize Levafix dye molecules, thermodynamic parametric data were quantified. Arrhenius formula was used to investigate the activation energy, $k_2 = Ae^{\frac{-E_a}{RT}}$, with the activation energy (E_a) of the Levafix dye oxidation being based on the second-order kinetic constant, where R is the gas constant (8.314 J mol^{-1}K^{-1}), T is the temperature in kelvin, and A is the pre-exponential factor that is considered to be constant with respect to temperature [32]. Taking the natural log of the Arrhenius formula yields the following:

$$\ln k_2 = \ln A - \frac{E_a}{RT} \tag{14}$$

Table 3. Fitted rate constants for the oxidation reaction of dye-containing wastewater *.

Kinetic Model	Parameter	Temperature, °C			
		25	40	50	60
Zero-order	k_0 (mg·min^{-1})	2.673	2.62	2.459	2.403
	$t_{1/2}$ (min)	72.171	70.794	66.393	64.881
	R^2	0.54	0.49	0.39	0.37
First-order	k_1 (min^{-1})	0.223	0.293	0.33	0.317
	$t_{1/2}$ (min)	3.107	2.365	2.100	2.186
	R^2	0.90	0.94	0.81	0.72
Second-order	k_2 (L·mg^{-1} min^{-1})	0.044	0.164	0.328	0.341
	$t_{1/2}$ (min)	0.4208	0.1129	0.0564	0.0543
	R^2	0.97	0.90	0.98	0.97

* k_0, k_1, k_2: kinetic rate constants of zero-, first-, and second-reaction kinetic models; C_o and C_t: dye concentrations at initial time and time t; t: time; R^2: correlation coefficient; $t_{1/2}$ half-life time.

A plot of $\ln k_2$ versus $1/T$ could be used to investigate the value of E_a (Figure 11). Figure 11 displays the relationship of the Levafix Dark Blue dye oxidation through the modified Fuller's earth-based photo Fenton system reaction. The value of E_a of the process was recorded to be 50.8 KJ mol^{-1}.

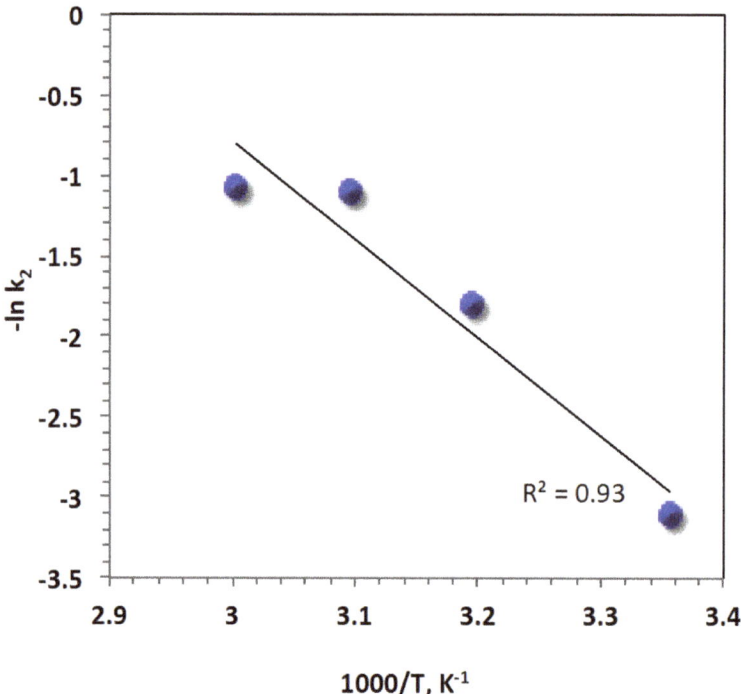

Figure 11. Plot of $\ln k_2$ versus $1000/T$ for the modified Fuller's earth Fenton system.

Other various thermodynamic variables, such as the enthalpy ($\Delta H'$), entropy ($\Delta S'$), and free energy ($\Delta G'$) of activation, were assessed utilizing Eyring's equation with the utilization of E_a and k_2 values according to the following relation [24]:

$$k_2 = \frac{k_B T}{h} e^{(-\frac{\Delta G'}{RT})} \tag{15}$$

where k_B is the Boltzmann constant, and h is Planck's constant. The thermodynamic parameters for Levafix dye oxidation were estimated accordingly, and they are displayed in Table 4.

Table 4. Thermodynamic properties of organics removal using modified Fuller's earth Fenton oxidation.

Parameters Thermodynamic Parameters	Temperature (°C)			
	25	40	50	60
$\Delta G'$ (kJ/mol)	80,720.78	81,487.94	82,314.45	84,839.69
$\Delta H'$ (kJ/mol)	−2426.76	−2551.47	−2634.61	−2717.75
$\Delta S'$ (J/mol)	−279.02	−268.49	−263.00	−262.93
Ea (kJ/mol)			50.8	

An investigation of the data in Table 4 found that the positive values of $\Delta H\prime$ across the studied temperature range indicate that the reaction is endothermic. Moreover, $\Delta G\prime$ exhibited positive values, which means that the process is non-spontaneous. This result might be because of the formation of a well-solvated structure between the dye molecules and the OH radical species. Moreover, the negative entropy values also support this.

3.2.6. Catalyst Stability and Reusability

To investigate the adequacy of the catalyst activity, its cyclic use was examined. To assist this investigation, the catalyst was collected after each use, washed with distilled water, and then subjected to oven drying for 1 h (105 °C). The catalyst loss was calculated, and it did not reach more than 1%. Then, the recovered catalyst was used for dye treatment at the optimal reagent doses of 818 mg/L, 1.02 mg/L, and 3.0 for H_2O_2, the Fuller's earth catalyst, and pH. The catalyst was repeatedly used, and its efficiency for treatment was examined. The catalyst reactivity reduced, achieving only 73% dye removal in the sixth cycle compared to the 99% achieved with the use of the fresh catalyst, as shown in Figure 12. However, it is worth mentioning that the catalyst was still efficient in removing dye from the aqueous solution. This reduction in the catalyst efficiency could be associated with the dye molecules occupying the active sites of the catalyst surface. Thus, its efficiency in producing hydroxyl radicals is reduced, and they are the horsepower of the oxidation reaction. Such a reduction in the catalyst efficacy after multiple uses has previously been reported in the literature [6] investigating the magnetized biomass catalyst recyclability in treating polluted wastewater.

It is worth mentioning that various researchers [6,11,33] have suggested that using a solvent for dye desorption might result in the regeneration of the catalyst. Moreover, eliminating the dye molecules occupying the catalyst surface through temperature elevation might also regenerate the catalyst. Additionally, changing the pH of the medium to achieve dye desorption could facilitate catalyst regeneration [34]. Moreover, introducing supercritical CO_2 to substitute the organic solvents for the desorption ability is a promising technique since it overcomes the environmental concerns regarding hazardous solvents. These suggested methods might elaborate the stability and recyclability of Fuller's earth. Not only can these methods regenerate the Fuller's earth material, but they can also recover and enable the collection of the dye adsorbate rather than destroying it, which is suggested to be an ideal sustainable solution. Therefore, these techniques are recommended for improving catalyst regeneration.

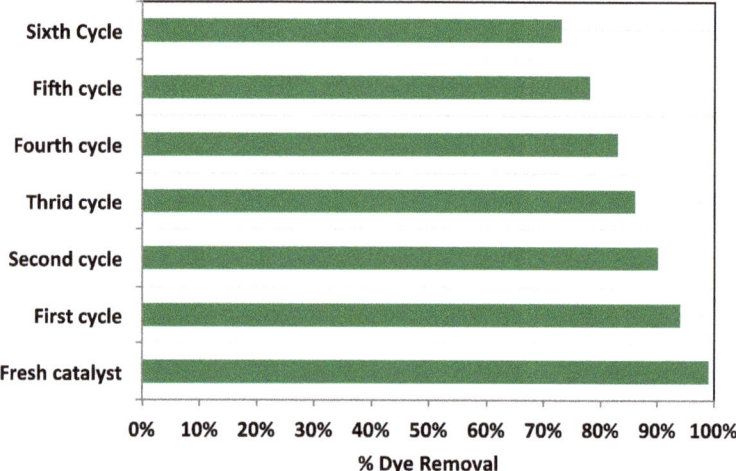

Figure 12. Catalyst reusability efficiency for cyclic use.

3.2.7. Data Comparison with Previous Fenton Studies

A comparison of various data presented in previous studies cited in the literature using the Fenton reaction to treat various types of dyes was carried out, and the results were compared with those of the current investigation. The comparison study is based on the modified Fenton system to examine the adequacy of the modified current study, and the comparison data are exhibited in Table 5. Regarding the oxidation removal efficiencies of the different processes, according to the data tabulated in Table 5, the Fenton reagent could achieve almost complete oxidation of various dye types. However, some systems are characterized by a low efficiency, such as the H_2O_2/Fe-zeolite and H_2O_2/Co-Fe_2O_4 systems. Additionally, it is worth mentioning that, in the current study of the modified Fenton system based on the use of Fullers' earth as a naturally occurring catalyst, only a limited time is needed in the current investigation for the reaction (20 min) compared to the other systems. Moreover, the current modified Fenton system is an efficient, superior mode of treatment since it is based on a naturally abundant material that is considered a minimal or costless catalyst. Thus, this current suggestive study is much better and cheaper, especially when the Fullers' earth catalyst is used as the source of the photo-Fenton system. Additionally, the process is environmentally friendly compared to the other techniques listed in Table 5, as they are not based on the use of naturally occurring substances as catalysts.

Although the other oxidation reactions also exhibited high dye removal efficiencies that almost resulted in complete removal, it is significant to mention that the catalyst source in the other studies was a metal-based synthesized catalyst. In the current study, the catalyst was based on a naturally abundant material. The efficiency of the current method being higher than that of the other Fenton technologies in Table 5 might also be associated with the occurrence of multiple metals in the catalyst, which increases the oxidation rate. Further, the catalyst dose in the current investigation was quite reasonable in comparison to that in the other studies, and the hydrogen peroxide reagent was not quite as high, with some studies reaching 2720 mg/L. However, in the other studies, only a minimal amount of hydrogen peroxide was essential, reaching 0.68 mg/L. Moreover, a higher reaction time was essential, reaching 150 min with a treatment efficiency of 79%. Additionally, it is worth mentioning that such data are attained when applying a costless, naturally abundant catalyst in treatment in comparison to the other mentioned systems in Table 5. Typically, the use of natural materials is superior in treatment due to their advantages of having environmentally benign characteristics.

Table 5. Comparing efficiency of various Fenton-based systems in oxidizing dyes.

Fenton System	Target Dye Pollutant	Initiation Source	Operating Conditions	Oxidation Time	Efficiency	Ref.
H_2O_2/Fullers' earth	Levafix blue	Ultraviolet illumination	Catalyst 1.02 g/L; H_2O_2 818 mg/L; pH 3.0; Temperature 298 K	20 min	99%	Current work
H_2O_2/Clay-Fe_3O_4	Levafix blue	Ultraviolet illumination	Catalyst 1 g/L; H_2O_2 400 mg/L; pH 3.0; Temperature 300 K	30 min	100%	[13]
H_2O_2/Fe-laponite	Acid Black 1	Ultraviolet illumination	Catalyst 1 g/L; H_2O_2 218 mg/L; pH 3.0; Temperature 298 K	120 min	100%	[34]
H_2O_2/Fe-zeolite	Acid Blue 74	Ultraviolet illumination	Catalyst 0.3 g/L; H_2O_2 228 mg/L; pH 5.0; Temperature 298 K	120 min	55%	[35]
H_2O_2/$Fe_{2.46}Ni_{0.54}O_4$	Methylene blue	Dark Fenton	Catalyst 1.0 g/L; H_2O_2 340 mg/L; pH 7.0; Temperature 298 K	50 min	10%	[36]
H_2O_2/Co-Fe_2O_4	Methylene blue	Natural solar radiation	Catalyst 0.2 g/L; H_2O_2 0.68 mg/L; pH 5.0; Temperature 298 K	150 min	79%	[35]
H_2O_2/Mn-Fe_2O_4	Methylene blue	Natural solar radiation	Catalyst 1.0 g/L; H_2O_2 2720 mg/L; pH 7.0; Temperature 298 K	80 min	100%	[36]
H_2O_2/Cu-Fe_2O_4	Methylene blue	Ultraviolet illumination	Catalyst 0.2 g/L; H_2O_2 0.68 mg/L; pH 5.0; Temperature 298 K	80 min	90%	[37]

3.2.8. Validation of Reclaimed Water

A physicochemical characterization of the water was carried out to determine its potential application. Both dyed wastewater effluents, as well as the water sample treated with the modified Fenton system were characterized. The median values of the wastewater properties, before and after treatment, were determined, and they are listed in Table 6. The performance of the modified Fenton system is clear from the data displayed in Table 6. The results demonstrate that, under UV illumination conditions using the Fenton catalyst, the Chemical Oxygen Demand (COD) reduced from 420 to 38 mg/L. The non-treated dye-containing wastewater presented a high concentration of COD. This resulted from effluent oxidation. Thus, the Fenton reagent treatment was able to oxidize and degrade most of the organics in the wastewater (90% COD removal). Moreover, when compared to the non-treated water, the reclaimed water, that is, the modified-Fenton-treated wastewater, exhibited lower dissolved oxygen (DO), turbidity, and suspended solids. However, the pH declined due to the Fenton reaction being carried out in the acidic range. It is worth mentioning that this acidic range might be altered prior to reuse. Such technology suggests that wastewater might be reused without severe toxicity accumulation. The quality of the dye wastewater treated using the modified Fenton treatment system is good enough for reuse in textile dyeing processes [38–40]. Hence, to verify the possibility of reusing reclaimed water, it is important to perform reuse tests.

Table 6. Wastewater characteristics, including physicochemical properties.

Parameter (Unit)	Suspended Solids (mg/L)	Turbidity (NTU)	pH	DO (mg/L)	COD (mg-$_{COD}$/L)
Wastewater	120	34	5.0	2.4	420
Treated water	38	2.5	3.0	2.3	28

4. Conclusions

In the current investigation, Levafix dye oxidation as an example of reactive dye molecules in aqueous media was conducted using a modified Fenton technique with Fuller's earth as the elemental catalyst source. SEM and FTIR suggested that the oxidation reaction was highly efficient due to the presence of a porous medium and elemental sources. The Levafix Dark Blue dye removal rates verified the effectiveness of the Fuller's earth-based Fenton treatment system. Response surface methodology based on a Box–Behnken factorial design was used to optimize the operating conditions, and the optimum values were 818 and 1.02 mg/L for Fuller's earth and hydrogen peroxide, respectively. This investigation found that using an optimal pH of 3.0 resulted in a 99% removal rate within 15 min of the illumination time. The kinetic rate equation for the system was best modeled by a second-order equation that well fitted the experimental results. The data of the experimental work, coupled with the calculated thermodynamic parameters, verified the significance of the system in removing azo dye. Therefore, this study demonstrates a sustainable and environmentally benign Fenton system since it is based on a naturally occurring catalyst. It is worth mentioning that the results of the current study have led to the application of such a technique in real textile wastewater effluent. However, more data are required to apply such a system to real textile effluent for both scientists and industrial operators. Additionally, much effort can be made to better design solar photocatalysis to satisfy the industrial sector and operate in a more economic flexible system to meet wastewater treatment requirements. Thus, upgrading such technology could highlight a new opportunity for the industrial sector.

Author Contributions: Conceptualization, M.A.T.; methodology, M.M.N. and M.A.T.; software, H.A.N.; formal analysis, H.A.N.; investigation, M.A.T.; data curation, H.A.N.; writing—original draft, M.M.N. and M.A.T.; writing—review and editing, M.A.T. and H.A.N.; project administration, M.M.N.; funding acquisition, M.M.N. All authors have read and agreed to the published version of the manuscript.

Funding: The authors extend their appreciation to Prince Sattam bin Abdulaziz University for funding this research work through project number (PSAU/2023/01/31544).

Institutional Review Board Statement: Not applicable.

Informed Consent Statement: Not applicable.

Data Availability Statement: Data are available upon request.

Acknowledgments: The authors extend their appreciation to Prince Sattam bin Abdulaziz University for funding this research work.

Conflicts of Interest: The authors declare no conflict of interest.

References

1. Shimi, A.K.; Parvathiraj, C.; Kumari, S.; Dalal, J.; Kumar, V.; Wabaidur, S.M.; Alothman, Z.A. Green synthesis of SrO nanoparticles using leaf extract of *Albizia julibrissin* and its recyclable photocatalytic activity: An eco-friendly approach for treatment of industrial wastewater. *Environ. Sci.* **2022**, *1*, 849–861. [CrossRef]
2. Uddin, F. Clays, nanoclays, and montmorillonite minerals. *Metall. Mater. Trans. A* **2008**, *39*, 2804–2814. [CrossRef]
3. Uddin, F. *Montmorillonite: An Introduction to Properties and Utilization*; IntechOpen: London, UK, 2018.
4. Rohilla, S.; Gupta, A.; Kumar, V.; Kumari, S.; Petru, M.; Amor, N.; Noman, M.T.; Dalal, J. Excellent UV-light triggered photocatalytic performance of ZnO.SiO$_2$ nanocomposite for water pollutant compound methyl orange dye. *Nanomaterials* **2021**, *11*, 2548. [CrossRef] [PubMed]
5. Safwat, S.M.; Medhat, M.; Abdel-Halim, H. Phenol adsorption onto kaolin and fuller's earth: A comparative study with bentonite. *Desalination Water Treat.* **2019**, *155*, 197–206. [CrossRef]
6. Nour, M.; Tony, M.A.; Nabawy, H. Immobilization of magnetic nanoparticles on cellulosic wooden sawdust for competitive Nudrin elimination from environmental waters as a green strategy: Box-Behnken Design optimization. *Int. J. Environ. Res. Public Health* **2022**, *19*, 15397. [CrossRef]
7. Lv, Q.; Li, G.; Sun, H.; Kong, L.; Lu, H.; Gao, X. Preparation of magnetic core/shell structured γ-Fe$_2$O$_3$@ Ti-tmSiO$_2$ and its application for the adsorption and degradation of dyes. *Microporous Mesoporous Mater.* **2014**, *186*, 7–13. [CrossRef]

8. Su, R.; Chai, L.; Tang, C.; Li, B.; Yang, Z. Comparison of the degradation of molecular and ionic ibuprofen in a UV/H_2O_2 system. *Water Sci. Technol.* **2018**, *77*, 2174–2183. [CrossRef]
9. Shah, J.; Jan, M.R.; Muhammad, M.; Ara, B.; Fahmeeda. Kinetic and equilibrium profile of the adsorptive removal of Acid Red 17 dye by surfactant-modified fuller's earth. *Water Sci. Technol.* **2017**, *75*, 1410–1420. [CrossRef]
10. Su, R.; Dai, X.; Wang, H.; Wang, Z.; Li, Z.; Chen, Y.; Luo, Y.; Ouyang, D. Metronidazole degradation by UV and UV/H_2O_2 advanced oxidation processes: Kinetics, mechanisms, and effects of natural water matrices. *Int. J. Environ. Res. Public Health* **2022**, *19*, 12354. [CrossRef]
11. Wang, Y.; Ye, X.; Chen, G.; Li, D.; Meng, S.; Chen, S. Synthesis of $BiPO_4$ by Crystallization and Hydroxylation with Boosted Photocatalytic Removal of Organic Pollutants in Air and Water. *J. Hazard. Mater.* **2020**, *399*, 122999. [CrossRef]
12. Ren, B.; Wang, T.; Qu, G.; Deng, F.; Liang, D.; Yang, W.; Liu, M. In Situ Synthesis of g-C_3N_4/TiO_2 Heterojunction Nanocomposites as a Highly Active Photocatalyst for the Degradation of Orange II under Visible Light Irradiation. *Environ. Sci. Pollut. Res.* **2018**, *25*, 19122–19133. [CrossRef]
13. Thabet, R.H.; Fouad, M.K.; El Sherbiny, S.A.; Tony, M.A. Identifying optimized conditions for developing dewatered alum sludge based photocatalyst to immobilize a wide range of dye contamination. *Appl. Water Sci.* **2022**, *12*, 210. [CrossRef]
14. Beltran-Pérez, O.D.; Hormaza-Anaguano, A.; Zuluaga-Diaz, B.; Cardona-Gallo, S.A. Structural modification of regenerated fuller earth and its application, in the adsorption of anionic and cationic dyes. *Dyna* **2015**, *82*, 165–171. [CrossRef]
15. Nguyen, T.T.; Huynh, K.A.; Padungthon, S.; Pranudta, A.; Amonpattaratkit, P.; Tran, L.B.; Phan, P.T.; Nguyen, N.H. Synthesis of natural flowerlike iron-alum oxide with special interaction of Fe-Si-Al oxides as an effective catalyst for heterogeneous Fenton process. *J. Environ. Chem. Eng.* **2021**, *9*, 105732. [CrossRef]
16. Mosallanejad, S.; Dlugogorski, B.Z.; Kennedy, E.M.; Stockenhuber, M. On the Chemistry of Iron Oxide Supported on γ-Alumina and Silica Catalysts. *ACS Omega* **2018**, *3*, 5362. [CrossRef]
17. He, D.; Zhang, C.; Zeng, G.; Yang, Y.; Huang, D.; Wang, L.; Wang, H. A Multifunctional Platform by Controlling of Carbon Nitride in the Core-Shell Structure: From Design to Construction, and Catalysis Applications. *Appl. Catal. B Environ.* **2019**, *258*, 117957. [CrossRef]
18. Tony, M.A. Valorization of undervalued aluminum-based waterworks sludge waste for the science of The 5 Rs' criteria. *Appl. Water Sci.* **2022**, *12*, 20. [CrossRef]
19. Wang, Z.; Wang, T.; Wang, Z.; Jin, Y. The adsorption and reaction of a titanate coupling reagent on the surfaces of different nanoparticles in supercritical CO_2. *J. Colloid. Intermol. Sci.* **2006**, *304*, 152. [CrossRef]
20. Pourali, P.; Behzad, M.; Arfaeinia, H.; Ahmadfazeli, A.; Afshin, S.; Poureshg, Y.; Rashtbari, Y. Removal of acid blue 113 from aqueous solutions using low-cost adsorbent: Adsorption isotherms, thermodynamics, kinetics and regeneration studies. *Sep. Sci. Technol.* **2021**, *18*, 3079. [CrossRef]
21. Raut-Jadhav, S.; Pinjari, D.V.; Saini, D.R.; Sonawane, S.H.; Pandit, A.B. Intensification of degradation of methomyl (carbamate group pesticide) by using the combination of ultrasonic cavitation and process intensifying additives. *Ultrason. Sonochemistry* **2016**, *31*, 135–142. [CrossRef]
22. Najjar, W.; Chirchi, L.; Santosb, E.; Ghorhel, A. Kinetic study of 2-nitrophenol photodegradation on Al-pillared montmorillonite doped with copper. *J. Environ. Monit.* **2001**, *3*, 697–701. [CrossRef] [PubMed]
23. Guan, S.; Yang, H.; Sun, X.; Xian, T. Preparation and promising application of novel $LaFeO_3$/BiOBr heterojunction photocatalysts for photocatalytic and photo-Fenton removal of dyes. *Opt. Mater.* **2020**, *100*, 109644. [CrossRef]
24. Tony, M.A.; Lin, L.S. Performance of acid mine drainage sludge as an innovative catalytic oxidation source for treating vehicle-washing wastewater. *J. Dispers. Sci. Technol.* **2020**, *43*, 50–60. [CrossRef]
25. Fang, P.; Wang, Z.; Wang, W. Enhanced Photocatalytic Performance of ZnTi-LDHs with Morphology Control. *CrystEngComm* **2019**, *21*, 7025–7031. [CrossRef]
26. Markandeya; Shukla, S.P.; Dhiman, N.; Mohan, D.; Kisku, G.C.; Roy, S. An efficient removal of disperse dye from wastewater using zeolite synthesized from cenospheres. *J. Hazard. Toxic Radioact. Waste* **2017**, *21*, 04017017. [CrossRef]
27. Rashed, M.N.; El Taher, M.A.D.; Fadlalla, S.M. Photocatalytic degradation of Rhodamine-B dye using composite prepared from drinking water treatment sludge and nano TiO_2. *Environ. Quality Manag.* **2022**, *31*, 175. [CrossRef]
28. Elsayed, S.A.; El-Sayed, E.; Tony, M. Impregnated chitin biopolymer with magnetic nanoparticles to immobilize dye from aqueous media as a simple, rapid and efficient composite photocatalyst. *Appl. Water Sci.* **2022**, *12*, 252. [CrossRef]
29. Nichela, D.A.; Berkovic, A.M.; Costante, M.R.; Juliarena, M.P.; Einschlag, F.S.G. Nitrobenzene degradation in Fenton-like systems using Cu (II) as catalyst. Comparison between Cu (II)- and Fe (III)-based systems. *Chem. Eng. J.* **2013**, *228*, 1148–1157. [CrossRef]
30. Buthiyappan, A.; Raman, A.A.; Daud, W.M.W. Development of an advanced chemical oxidation wastewater treatment system for the batik industry in Malaysia. *RSC Adv.* **2016**, *6*, 25222. [CrossRef]
31. Lopez-Lopez, C.; Martín-Pascual, J.; Martínez-Toledo, M.V.; González-López, J.; Hontoria, E.; Poyatos, J.M. Effect of the operative variables on the treatment of wastewater polluted with phthalo blue by H_2O_2/UV process. *Water Air Soil Pollut.* **2013**, *224*, 1725. [CrossRef]
32. Santana, C.S.; Ramos, M.D.N.; Velloso, C.C.V.; Aguiar, A. Kinetic evaluation of dye decolorization by Fenton processes in the presence of 3-hydroxyanthranilic acid. *Int. J. Environ. Res. Public Health* **2019**, *16*, 1602. [CrossRef]
33. Feng, J.; Hu, X.; Yue, P.L.; Zhu, H.Y.; Lu, G.Q. Discoloration and mineralization of Reactive Red HE-3B by heterogeneous photo-Fenton reaction. *Water Res.* **2003**, *37*, 3776. [CrossRef]

34. Parker, H.L.; Budarin, V.L.; Clark, J.H.; Hunt, A.J. Use of Starbon for the Adsorption and Desorption of Phenols. *ACS Sustain. Chem. Eng.* **2013**, *1*, 1311–1318. [CrossRef]
35. Sum, O.S.; Feng, J.; Hu, X.; Yue, P.L. Pillared laponite clay-based Fe nanocomposites as heterogeneous catalysts for photo-Fenton degradation of acid black. *Chem. Eng. Sci.* **2004**, *59*, 5269.
36. Kalam, A.; Al-Sehemi, A.G.; Assiri, M.; Du, G.; Ahmad, T.; Ahmad, I.; Pannipara, M. Modified solvothermal synthesis of cobalt ferrite ($CoFe_2O_4$) magnetic nanoparticles photocatalysts for degradation of methylene blue with H_2O_2/visible light. *Results Phys.* **2018**, *8*, 1046–1053. [CrossRef]
37. Desai, H.B.; Hathiya, L.; Joshi, H.; Tanna, A. Synthesis and characterization of photocatalytic $MnFe_2O_4$ nanoparticles. *Mater. Today Proc.* **2020**, *21*, 1905–1910. [CrossRef]
38. Guo, X.; Wang, K.; Xu, Y. Tartaric acid enhanced $CuFe_2O_4$-catalyzed heterogeneous photo-Fenton-like degradation of methylene blue. *Mater. Sci. Eng.* **2019**, *245*, 75. [CrossRef]
39. Meghwal, K.; Agrawal, R.; Kumawat, S.; Jangid, N.K.; Ameta, C. Chemical and Biological Treatment of Dyes. In *Impact of Textile Dyes on Public Health and the Environment*; IGI Global: Hershey, PA, USA, 2020. [CrossRef]
40. Meghwal, K.; Kumawat, S.; Ameta, C.; Jangid, N.K. Effect of dyes on water chemistry, soil quality, and biological properties of water. In *Impact of Textile Dyes on Public Health and the Environment*; IGI Global: Hershey, PA, USA, 2020; pp. 90–114. [CrossRef]

Disclaimer/Publisher's Note: The statements, opinions and data contained in all publications are solely those of the individual author(s) and contributor(s) and not of MDPI and/or the editor(s). MDPI and/or the editor(s) disclaim responsibility for any injury to people or property resulting from any ideas, methods, instructions or products referred to in the content.

MDPI
St. Alban-Anlage 66
4052 Basel
Switzerland
www.mdpi.com

Applied Sciences Editorial Office
E-mail: applsci@mdpi.com
www.mdpi.com/journal/applsci

Disclaimer/Publisher's Note: The statements, opinions and data contained in all publications are solely those of the individual author(s) and contributor(s) and not of MDPI and/or the editor(s). MDPI and/or the editor(s) disclaim responsibility for any injury to people or property resulting from any ideas, methods, instructions or products referred to in the content.

www.ingramcontent.com/pod-product-compliance
Lightning Source LLC
LaVergne TN
LVHW070605100526
838202LV00012B/568